THE AIR & ITS WAYS

THE AIR & ITS WAYS

THE REDE LECTURE (1921) IN THE UNIVERSITY
OF CAMBRIDGE, WITH OTHER CONTRIBUTIONS
TO METEOROLOGY FOR SCHOOLS AND COLLEGES

BY

SIR NAPIER SHAW, Sc.D., F.R.S.

COMMANDER OF THE ORDER OF S. TIAGO DA ESPADA OF PORTUGAL;
PROFESSOR OF METEOROLOGY, ROYAL COLLEGE OF SCIENCE;
READER IN METEOROLOGY IN THE UNIVERSITY OF LONDON;
LATE DIRECTOR OF THE METEOROLOGICAL OFFICE; HONORARY
FELLOW OF EMMANUEL COLLEGE; HONORARY DOCTOR OF LAWS
OF THE UNIVERSITIES OF ABERDEEN AND EDINBURGH; HONORARY
DOCTOR OF SCIENCE OF THE UNIVERSITIES OF DUBLIN AND
MANCHESTER AND OF HARVARD UNIVERSITY, CAMBRIDGE, MASS.;
HONORARY FOREIGN MEMBER OF THE AMERICAN ACADEMY
OF ARTS AND SCIENCES, BOSTON, AND OF THE
REALE ACCADEMIA DEI LINCEI, ROME

CAMBRIDGE
AT THE UNIVERSITY PRESS
1923

CAMBRIDGE
UNIVERSITY PRESS

University Printing House, Cambridge CB2 8BS, United Kingdom

Cambridge University Press is part of the University of Cambridge.

It furthers the University's mission by disseminating knowledge in the pursuit of education, learning and research at the highest international levels of excellence.

www.cambridge.org
Information on this title: www.cambridge.org/9781107511491

© Cambridge University Press 1923

First published 1923
First paperback edition 2015

A catalogue record for this publication is available from the British Library

ISBN 978-1-107-51149-1 Paperback

PREFACE

FROM time to time I have been asked by students who are also school-masters or mistresses whether I consider the commonly received explanation on physical principles of the general circulation of the atmosphere, as it has appeared for many years in text-books of physical geography, to be a satisfactory representation of the state of our knowledge of the atmosphere in the twentieth century. Hitherto my reply could only be that I had said what seemed to me to be worth saying on the subject in various lectures, and that some day if possible I would offer the text of the lectures as the best answer which I could give to the question.

Accordingly I have put together a number of lectures and essays that dealt with the physical explanation of the atmospheric circulation and to those I have added some papers designed to set out the present position of the application of meteorology to agriculture: a subject, like that of the general circulation of the atmosphere, of perennial interest not only to meteorologists but to all who have to think about the air and its ways.

Lectures and addresses on meteorological subjects are always easy to make and sometimes interesting to hear. Curiosity about the subject is universal and spontaneous. It requires little adroitness on the part of the lecturer to stimulate the curiosity about the mutual connexion of events, which we call "explanations." Illustrations abound and they are easily expressed in maps and photographs. In the course of the daily work at the Meteorological Office they accumulated faster than the opportunities for displaying them. But it requires a good deal of courage to take the next step and publish a collection of lectures on meteorological subjects, because the illustrations, which are the lecturer's fairy godmother, are the publisher's *bête-noire* and the author's despair.

There has been, therefore, a good deal of hesitation behind the production of this volume. The text is necessarily woven round illustrations which are used over and over again; prudence demands that publication should be deferred until there is some reasonable balance between text and pictures. It is hoped that the accumulation in this book will satisfy that condition. It is not a text-book: it shows the subject of meteorology in its work-day clothes with loose or missing buttons here and there, and the tailoring not always perfect. But for the workman a lecture is a useful opportunity of "trying on" and I hope the reader will accept this collection of essays with the amiability that becomes that operation.

My collection is not by any means homogeneous. It rambles from climatology to physics and dynamics and back again to agriculture; the main current of such thought as there is in it is the bringing of the ascertained and coordinated facts about the weather into relation with each other and thereby with the laws and principles of physics. In this connexion three new meteorological principles are introduced to the reader as inductively justified. First, the motion of air under balanced forces as a more effective representation of actual conditions than any which we can substitute for it by observation or by theory. Secondly, the "eviction" of air by the turbulent motion between rising air and its environment as an inevitable concomitant of convexion. Thirdly, the effective stratification of successive layers of air in consequence of the "resilience" which they owe to the excess of their temperature over that which corresponds with unlimited miscibility, and the limitations of convexion which are due thereto. This principle, which carries some sort of analogy to the effect of steel rods in "reinforcing" concrete, involves some recondite ideas about potential temperature or entropy which the gentle reader must be good enough to think about as part of the natural order.

These three principles come into various theoretical and practical considerations. The introduction of the first in 1913, so far as my own experience is concerned, has created little less than a revolution in my meteorological world. Things live and move; previously they were tethered by an indefinite attachment, and paralysed by an unknown quantity which blocked every avenue: it is the dominant note of the Halley Lecture of 1918. The second is, so far as I know, entirely new and has already opened up new possibilities. The third simply brings into definite form ideas which have been drifting loosely in meteorological literature for many years. Its definition will I hope lead in course of time to satisfactory evidence for the process of development of high pressures the formation of which has hitherto been unexplored.

The remaining essays are less ambitious from the scientific point of view and are intended to represent certain aspects of modern meteorology as a subject of interest to the general reader. But another new principle is visible in the work on wheat-crops, namely that plants and the soil in which they grow have long memories and the resulting crop depends upon the conditions of long ago as well as those of yesterday.

In the preparation of the work and of the original lectures upon which it is based I owe much to the assistance of Miss E. E. Austin, of Newnham College, who was my personal assistant for two years at the Meteorological Office, and has been seconded by the Air Ministry to help me in my work at the Imperial College.

I have also to record my thanks to the Controller of H.M. Stationery Office for permission to reproduce Figs. 1, 2, 22, 33, 39, 58–60, 74 and 97, taken from official publications of the Meteorological Office, and to the Director of the Meteorological Office for the loan of blocks for Figs. 1, 2, 23, 66 and 74. The material for Figs. 34 (a) and (b) is also derived from official publications of the Meteorological Office.

Thanks are also due to the Royal Meteorological Society for permission to reprint Papers 1 and 3 from the *Quarterly Journal*, and for the loan of blocks for Figs. 3, 4, 5 and 65; to Captain C. J. P. Cave for the original negatives or prints of the cloud photographs reproduced in Figs. 6–12, and for the loan of blocks for Figs. 99 and 100; to Major G. M. B. Dobson for permission to reproduce Fig. 65; to the proprietors of *The Times* for permission to reprint the paper on "The Drought of 1921"; and to the Mount Everest Committee for permission to reproduce the photograph of Mount Everest in Fig. 79.

Some maps and other original illustrations are common to this publication and also to the *Dictionary of Applied Physics* edited by

Sir Richard Glazebrook, published by Macmillan and Co., Ltd., where they appear in the article on "Thermodynamics of the Atmosphere."

The maps of mean rainfall (Figs. 16, 17 and 71), mean temperature (Figs. 18–21), mean cloudiness (Figs. 69, 70) have been taken from those compiled by Mr C. E. P. Brooks, with the assistance of the division of the Meteorological Office for the Réseau Mondial, for' the *Manual of Meteorology* which is still in hand; the maps of dew-point (Figs. 51–54) are drawn from data extracted by that division.

For Fig. 13, Contessa del Vento, I owe my thanks to the cloud-atlas of Signor L. Taffara of the R. Ufficio Centrale di Meteorologia e Geodinamica.

For the reading of the proofs my thanks are due also to Captain D. Brunt of the Meteorological Office, and I carry forward a sense of personal obligation to Mr J. B. Peace, a College friend of long standing, for the trouble which he has taken in arranging the material here presented; and to the staff of the Cambridge University Press for the skill and patience by which they have succeeded in making a book out of a collection of fragments.

NAPIER SHAW

SCHOOL OF METEOROLOGY,
IMPERIAL COLLEGE OF SCIENCE AND TECHNOLOGY,
June, 1922.

POSTSCRIPT

On the evening of January 30, 1923, while my story of *The Air and its Ways* was in the press, Mr Peace's friendly guidance was suddenly terminated by his untimely death. The result of our joint work, which was intended to be a preliminary example, now becomes a grateful memory of the helpfulness which was characteristic of his life.

N. S.

February 23, 1923

CONTENTS

ILLUSTRATIONS

THE CYCLONIC DEPRESSIONS OF MIDDLE LATITUDES (*contd.*)

THE NORMAL DISTRIBUTION OF THE METEORO-LOGICAL ELEMENTS OVER THE GLOBE

PLATES I–XXIV

The Plates which are inserted here are representative examples of the distribution of the meteorological elements over the globe, to which reference is made in many places in illustration of the lectures and essays which they precede. Together with the winds of Figs. 72 and 73 and the vertical sectional diagrams represented in Figs. 47, 55, 56, 57 and 68, they are working diagrams of the normal state of the atmosphere for the student of meteorology. In the standard meteorological scheme the month is the unit of time, and, for the work of a student, charts for each month are necessary. Those which appear here are selected as specimens from a series prepared for Part I of the *Manual of Meteorology*.

It must be remarked that the representation of the earth's surface by two circles, which is employed here, is not in accordance with any rule of geometrical projection, and the result ought therefore to be called a diagram and not a map. The method has its disadvantages but in the present examples they are less conspicuous than its advantages.

Plates I to III	represent the distribution of rainfall.			
„ IV to VII	„	„	„	temperature.
„ VIII to XI	„	„	„	dew-point.
„ XII and XIII	„	„	„	cloud.
„ XIV to XVII	„	„	„	pressure at 4000 metres.
„ XVIII⎱				⎰pressure at 8000 metres.
„ XIX ⎰	„	„	„	⎱ „ of the lower stratum.
„ XX to XXIII	„	„	„	pressure at sea-level.
„ XXIV	represents the normal resultant-flow of air.			

*** The "M.O. compilation" upon which the lines of the chart on the opposite page are based is that which is referred to on p. viii of the preface to this volume as carried out in the Meteorological Office for Part I of the *Manual of Meteorology*. The same abbreviation is used in quoting the authority for other charts in the same category.

JULY

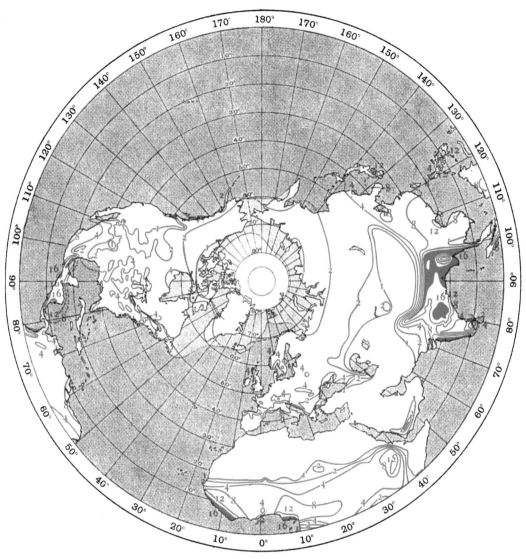

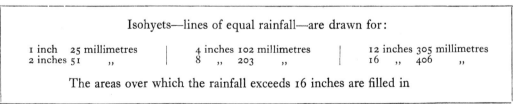

Isohyets—lines of equal rainfall—are drawn for:

| 1 inch | 25 millimetres | 4 inches | 102 millimetres | 12 inches | 305 millimetres |
| 2 inches | 51 ,, | 8 ,, | 203 ,, | 16 ,, | 406 ,, |

The areas over which the rainfall exceeds 16 inches are filled in

Fig. 71. Normal rainfall in July over the northern hemisphere: in inches.

PLATE II
Reference to text, p. 37.

ANNUAL RAINFALL IN THE NORTHERN HEMISPHERE
Authority: for the land, M.O. compilation; for the sea, Prof. Supan.

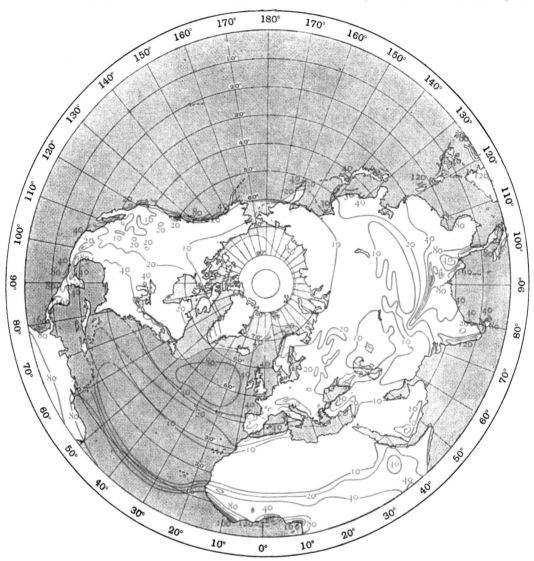

Isohyets—lines of equal rainfall—are drawn for:

10 inches 254 millimetres	40 inches 1016 millimetres	120 inches 3048 millimetres
20 ,, 508 ,,	80 ,, 2032 ,,	160 ,, 4064 ,,

The areas over which the rainfall exceeds 160 inches are filled in

Fig. 16. Normal rainfall for the year over the northern hemisphere: in inches.

ANNUAL RAINFALL IN THE SOUTHERN HEMISPHERE

Authority: for the land, M.O. compilation; for the sea, Prof. Supan.

PLATE III

Reference to text, p. 37.

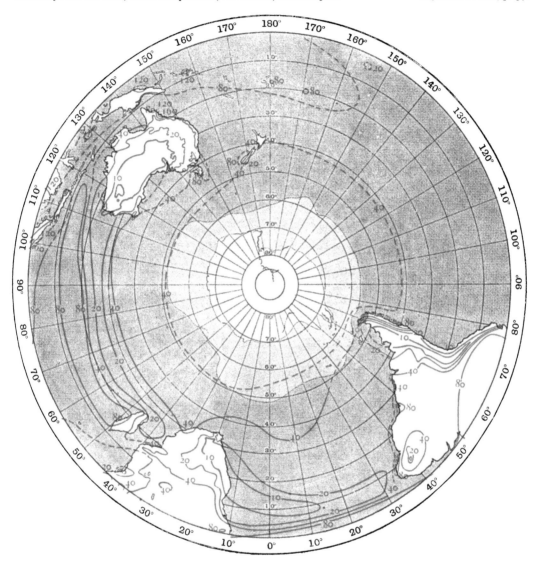

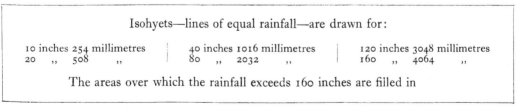

Isohyets—lines of equal rainfall—are drawn for:

| 10 inches 254 millimetres | 40 inches 1016 millimetres | 120 inches 3048 millimetres |
| 20 ,,   508   ,, | 80 ,,   2032   ,, | 160 ,,   4064   ,, |

The areas over which the rainfall exceeds 160 inches are filled in

Fig. 17. Normal rainfall for the year over the southern hemisphere: in inches.

JANUARY

Reference to text, pp. 37, 120, 141.   *Authority:* M.O. compilation.

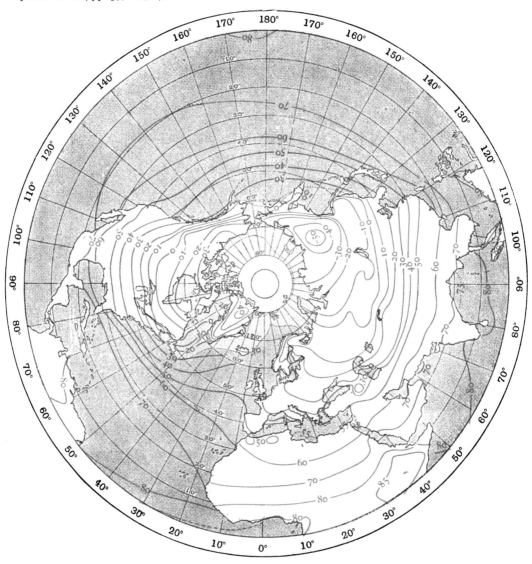

Equivalents below 30° F				The more lightly shaded area in the polar regions marks the probable range of ice between summer and winter.	Equivalents above 30° F			
°F	a	°F	a		°F	a	°F	a
− 50	227·4	− 10	249·7	Isotherms over land are plotted independently of those over sea. Temperatures over land have been reduced to sea-level.	30	271·9	60	288·6
− 40	233·0	0	255·2		32	273·0	70	294·1
− 30	238·6	10	260·8		40	277·4	80	299·7
− 20	244·1	20	266·3		50	283·0	90	305·2

The run of the isotherm of 80° in the equatorial region of the Pacific Ocean is not yet fully ascertained.

Fig. 18. Normal mean temperature of the air over the northern hemisphere in January: isotherms for steps of ten degrees of the Fahrenheit scale.

JULY

MEAN TEMPERATURE OF THE AIR AT SEA-LEVEL, NORTHERN HEMISPHERE   PLATE V

Authority: M.O. compilation.   *Reference to text*, pp. 37, 120, 141.

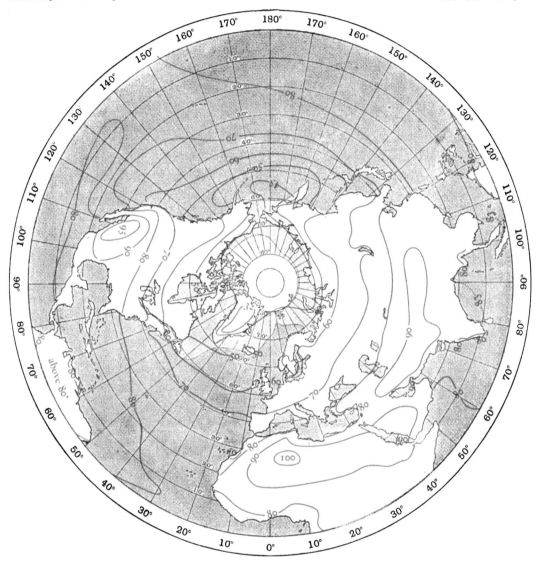

Equivalents				The unshaded area round the north pole marks the probable limit of field-ice in the northern summer. Isotherms over the sea are plotted independently of those over the land. Temperatures over land have been reduced to sea-level.	Equivalents			
°F	a	°F	a		°F	a	°F	a
32	273·0	50	283·0		70	294·1	90	305·2
40	277·4	60	288·6		80	299·7	100	310·8

Fig. 19. Normal mean temperature of the air over the northern hemisphere in July: isotherms for steps of ten degrees of the Fahrenheit scale.

JANUARY

Reference to text, pp. 37, 120, 141.   *Authority:* M.O. compilation.

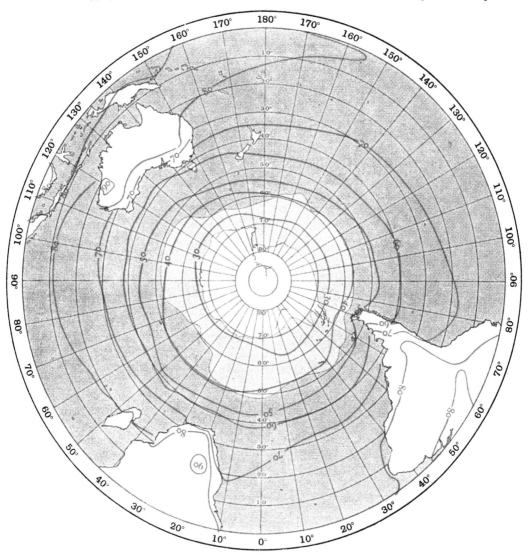

Equivalents				The unshaded area round the south pole marks the probable limit of field-ice in the southern summer.	Equivalents			
°F	a	°F	a	Isotherms over the sea are plotted independently of those over the land. Temperatures over land have been reduced to sea-level.	°F	a	°F	a
30	271·9	40	277·4		60	288·6	80	299·7
32	273·0	50	283·0		70	294·1	90	305·2

The run of the isotherms in the equatorial region of the Pacific Ocean is not yet fully ascertained.

Fig. 20. Normal mean temperature of the air over the southern hemisphere in January: isotherms for steps of ten degrees of the Fahrenheit scale.

JULY

MEAN TEMPERATURE OF THE AIR AT SEA-LEVEL, SOUTHERN HEMISPHERE   PLATE VII

Authority: M.O. compilation.   *Reference to text,* pp. 37, 120, 141.

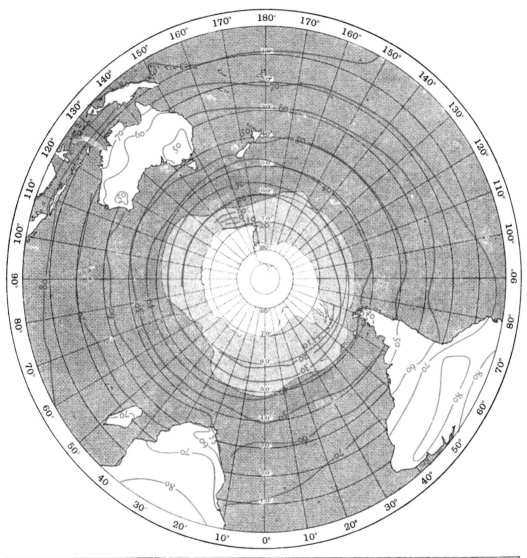

Equivalents				The more lightly shaded area in the polar region marks the probable range of the polar ice between summer and winter.	Equivalents			
°F	a	°F	a		°F	a	°F	a
– 20	244·1	20	266·3	Isotherms over the sea are plotted independently of those over the land. Temperatures over the land are reduced to sea-level.	40	277·4	60	288·6
– 10	249·7	30	271·9		50	283·0	70	294·1
0	255·2	32	273·0		55	285·8	80	299·7
10	260·8	35	274·7					

Fig. 21. Normal mean temperature of the air over the southern hemisphere in July: isotherms generally for steps of ten degrees of the Fahrenheit scale.

JANUARY

PLATE VIII   MEAN DEW-POINT OF THE AIR AT SEA-LEVEL, NORTHERN HEMISPHERE

Reference to text, pp. 120, 141.   *Authority:* Original compilation.

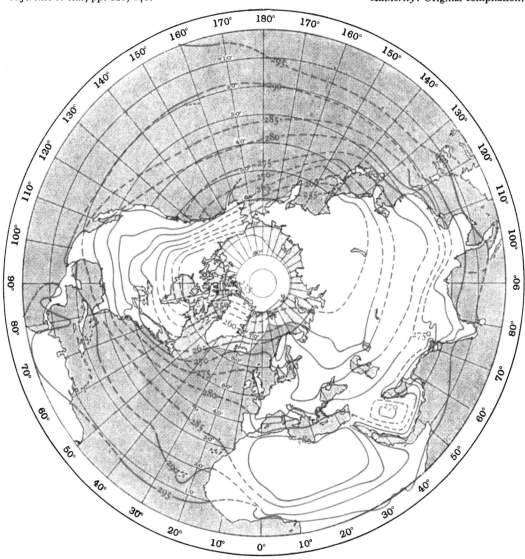

Equivalent pressure of aqueous vapour and of density at saturation			The mean dew-point has been obtained from the mean vapour-pressure at the several observatories and stations after reduction to sea-level by the formula:	Equivalent pressure of aqueous vapour and of density at saturation		
Dew-point	Vapour-pressure	Vapour-density		Dew-point	Vapour-pressure	Vapour-density
a	mb	g/m³		a	mb	g/m³
235	0·16	0·15		270	4·8	3·9
240	0·28	0·25	$$e_0 = e_h \,(1 + \cdot0004h),$$	273	6·1	4·9
245	0·47	0·42		275	7·1	5·6
250	0·78	0·67	where h is the height of the observatory	280	10·0	7·8
255	1·27	1·08	in metres.	285	14·1	10·7
260	1·99	1·66		290	19·4	14·5
265	3·10	2·54		295	26·5	19·5

Fig. 51. Normal mean temperature of saturation, dew-point or cloud-temperature of the air at sea-level over the northern hemisphere in January.

JULY

MEAN DEW-POINT OF THE AIR AT SEA-LEVEL, NORTHERN HEMISPHERE   PLATE IX

Authority: Original compilation.   *Reference to text,* pp. 120, 141.

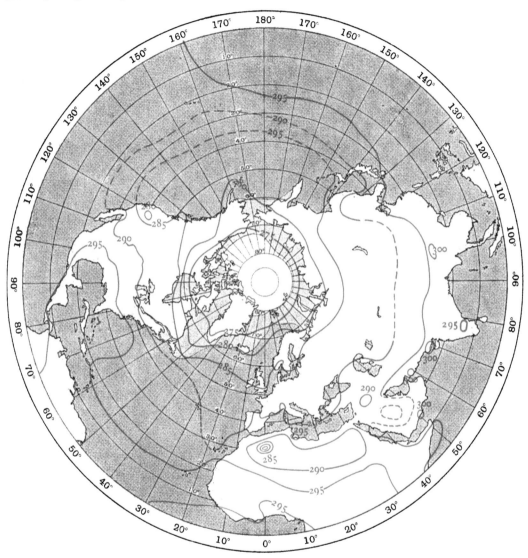

Equivalent pressure of aqueous vapour and of density at saturation			The mean dew-point has been obtained from the mean vapour-pressure at the several observatories and stations after reduction to sea-level by the formula:	Equivalent pressure of aqueous vapour and of density at saturation		
Dew-point	Vapour-pressure	Vapour-density		Dew-point	Vapour-pressure	Vapour-density
a	mb	g/m³	$e_0 = e_h\,(1 + \cdot0004h),$	a	mb	g/m³
275	7·1	5·6	where *h* is the height of the observatory in	290	19·4	14·5
280	10·0	7·8	metres.	295	26·5	19·5
285	14·1	10·7		300	35·7	25·8

Fig. 52. Normal mean temperature of saturation, dew-point or cloud-temperature of the air at sea-level over the northern hemisphere in July.

PLATE X MEAN DEW-POINT OF THE AIR AT SEA-LEVEL, SOUTHERN HEMISPHERE

Reference to text, pp. 120, 141. *Authority:* Original compilation.

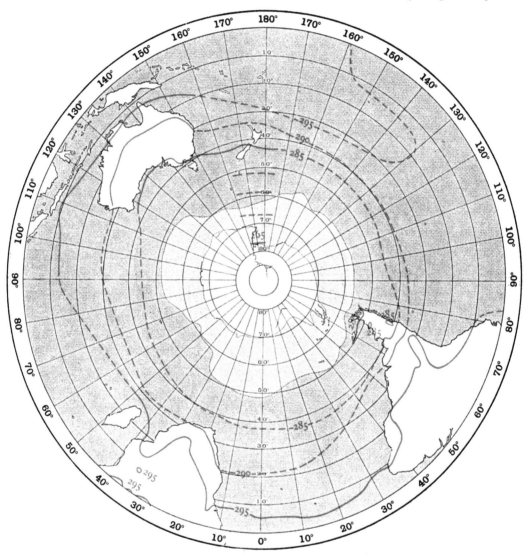

Equivalent pressure of aqueous vapour and of density at saturation			For reduction to sea-level see Plates VIII and IX.	Equivalent pressure of aqueous vapour and of density at saturation		
Dew-point	Vapour pressure	Vapour-density		Dew-point	Vapour pressure	Vapour-density
a	mb	g/m³	In explanation of the run of the line in the Western Pacific see Plate VI and the map of sea-temperatures for February, *Barometer Manual,* 1919 (Plate XIII).	a	mb	g/m³
265	3·1	2 5		285	14·1	10·7
270	4·8	3·9		290	19·4	14·5
275	7·1	5·6		295	26·5	19·5
280	10·0	7·8		(300)	35·7	25·8

Fig. 53. Normal mean temperature of saturation, dew-point or cloud-temperature of the air at sea-level over the southern hemisphere in January.

MEAN DEW-POINT OF THE AIR AT SEA-LEVEL, SOUTHERN HEMISPHERE   PLATE XI

Authority: Original compilation.   *Reference to text*, pp. 120, 141.

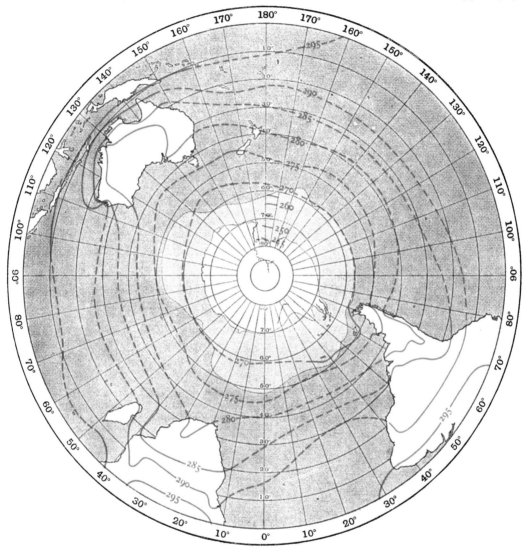

Equivalent pressure of aqueous vapour and of density at saturation			For reduction to sea-level see Plates VIII and IX.	Equivalent pressure of aqueous vapour and of density at saturation		
Dew-point	Vapour-pressure	Vapour-density		Dew-point	Vapour-pressure	Vapour-density
a	mb	g/m³		a	mb	g/m³
245	0·47	0·42		275	7·1	5·6
250	0·78	0·67		280	10·0	7·8
255	1·27	1·08		285	14·1	10·7
260	1·99	1·66		290	19·4	14·5
265	3·10	2·54		295	26·5	19·5

Fig. 54. Normal mean temperature of saturation, dew-point or cloud-temperature of the air at sea-level over the southern hemisphere in July.

JULY

PLATE XII   NORMAL DISTRIBUTION OF CLOUD IN THE NORTHERN HEMISPHERE

Reference to text, p. 143.   *Authority:* M.O. compilation.

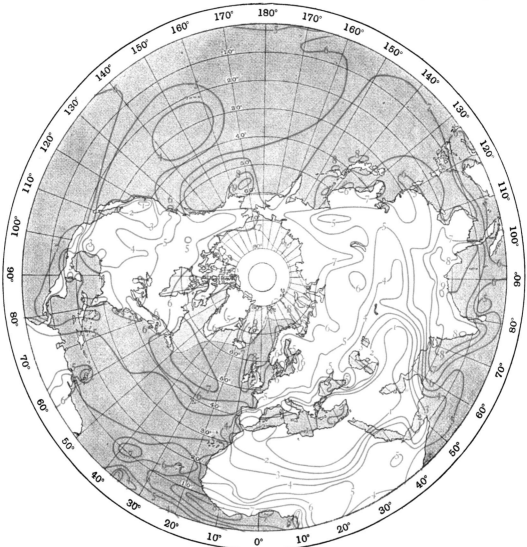

The lines are drawn from observations at fixed hours to represent the mean cloudiness for the whole month. They have been modified in consideration of additional information while this volume has been passing through the press and are liable to further modification when additional observations are incorporated.

Fig. 69. Isonephs: lines showing the number of tenths of the sky covered by cloud in the northern hemisphere in July.

JULY

NORMAL DISTRIBUTION OF CLOUD IN THE SOUTHERN HEMISPHERE   PLATE XIII

Authority: M.O. compilation.   *Reference to text*, p. 143.

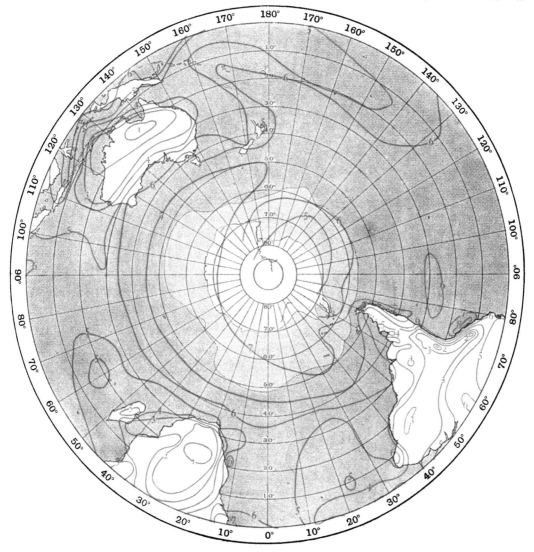

The lines are drawn from observations at fixed hours to represent the mean cloudiness for the whole month. They have been modified in consideration of additional information while this volume has been passing through the press and are liable to further modification when additional observations are incorporated.

Fig. 70. Isonephs: lines showing the number of tenths of the sky covered by cloud in the southern hemisphere in July.

PLATE XIV   NORMAL PRESSURE AT 4000 METRES, NORTHERN HEMISPHERE

Reference to text, pp. 40, 58, 72, 76.   *Authority:* L. Teisserenc de Bort.

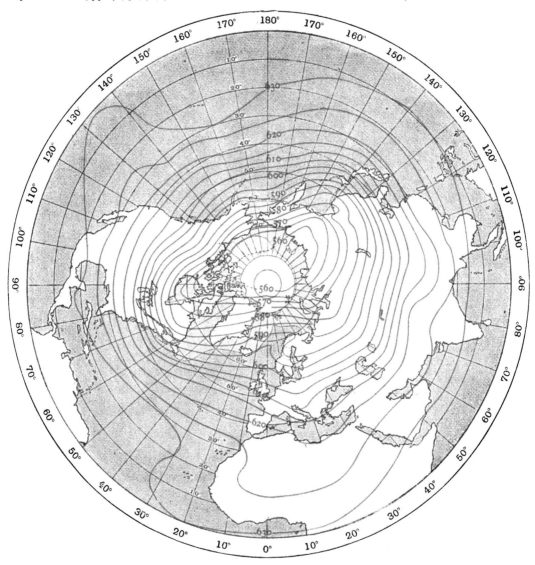

Equivalents			The pressures were obtained from the distribution at the surface by subtracting the pressure of the lower layer computed from the pressure and temperature at the surface. *Étude sur la circulation générale de l'atmosphère, Annales du Bureau Central Météorologique*, IV, 1887, pp. 35–44.	Equivalents		
mb	mm	in		mb	mm	in
595	446·3	17·571		575	431·3	16·980
590	442·5	17·423		570	427·5	16·833
585	438·8	17·276		565	423·8	16·685
580	435·0	17·128		560	420·0	16·537

The isobars are transformed from the originals in steps of four millimetres.

Fig. 24. Isobars in steps of five millibars at the level of 4000 metres in the northern hemisphere in January.

NORMAL PRESSURE AT 4000 METRES, NORTHERN HEMISPHERE   PLATE XV

Authority: L. Teisserenc de Bort.   *Reference to text,* pp. 40, 58, 76.

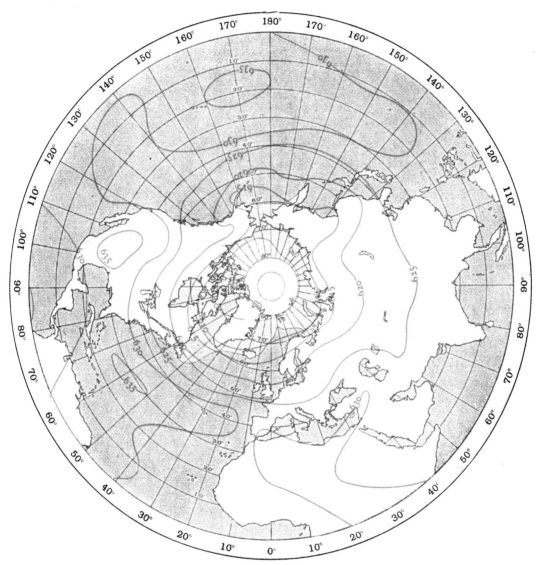

Equivalents			The pressures were obtained from the distribution at the surface by subtracting the pressure of the lower layer computed from the pressure and temperature at the surface. Étude sur la circulation générale de l'atmosphère, *Annales du Bureau Central Météorologique,* IV, 1887, pp. 35–44.	Equivalents		
mb	mm	in		mb	mm	in
635	476·3	18·752		615	461·3	18·162
630	472·5	18·605		610	457·5	18·014
625	468·8	18·457		605	453·8	17·866
620	465·0	18·309		600	450·0	17·719

The isobars are transformed from the originals in steps of four millimetres.

Fig. 25. Isobars in steps of five millibars at the level of 4000 metres in the northern hemisphere in July. Isobars for the level of 8000 metres are given in Plate XVIII.

JANUARY

PLATE XVI   NORMAL PRESSURE AT 4000 METRES, SOUTHERN HEMISPHERE

Reference to text, pp. 40, 58, 76.   *Authority:* L. Teisserenc de Bort.

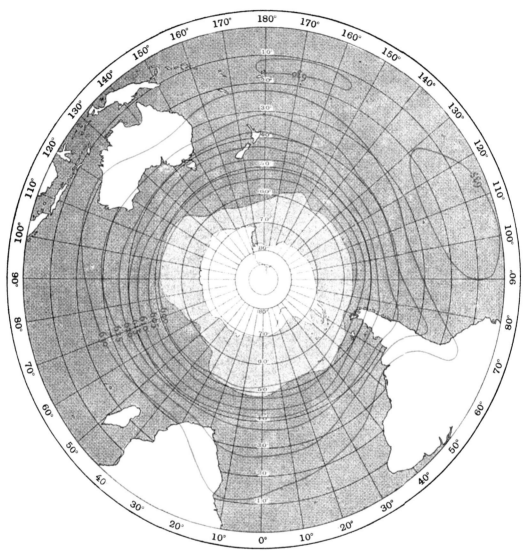

Equivalents			Computed from charts of the distribution expressed in millimetres of mercury, *Annales du Bureau Central Météorologique*, IV, 1887, pp. 35–44.	Equivalents		
mb	mm	in		mb	mm	in
635	476·3	18·752		615	461·3	18·162
630	472·5	18·605		610	457·5	18·014
625	468·8	18·457		605	453·8	17·866
620	465·0	18·309		600	450·0	17·719

Fig. 26. Isobars in steps of five millibars at the level of 4000 metres in the southern hemisphere in January.

JULY

NORMAL PRESSURE AT 4000 METRES, SOUTHERN HEMISPHERE

PLATE XVII

Authority: L. Teisserenc de Bort.

Reference to text, pp. 40, 58, 76.

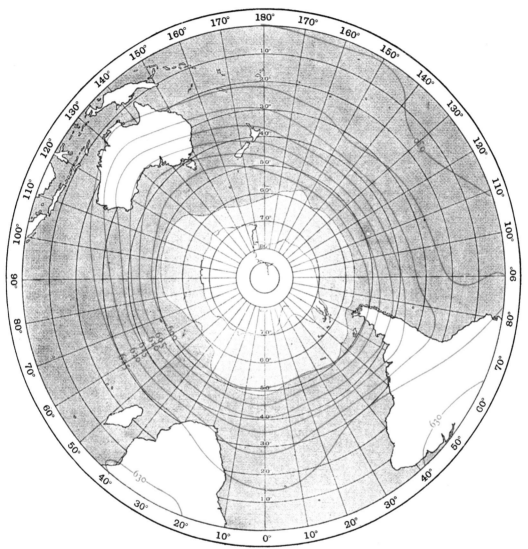

Equivalents			Computed from charts of the distri-bution expressed in millimetres of mercury, *Annales du Bureau Central Météorologique*, IV, 1887, pp. 35–44.	Equivalents		
mb	mm	in		mb	mm	in
635	476·3	18·752		615	461·3	18·162
630	472·5	18·605		610	457·5	18·014
625	468·8	18·457		605	453·8	17·866
620	465·0	18·309		600	450·0	17·719

Fig. 27. Isobars in steps of five millibars at the level of 4000 metres in the southern hemisphere in July.

JULY

PLATE XVIII   NORMAL PRESSURE AT 8000 METRES, NORTHERN HEMISPHERE

Reference to text, p. 130.   *Authority:* Original.

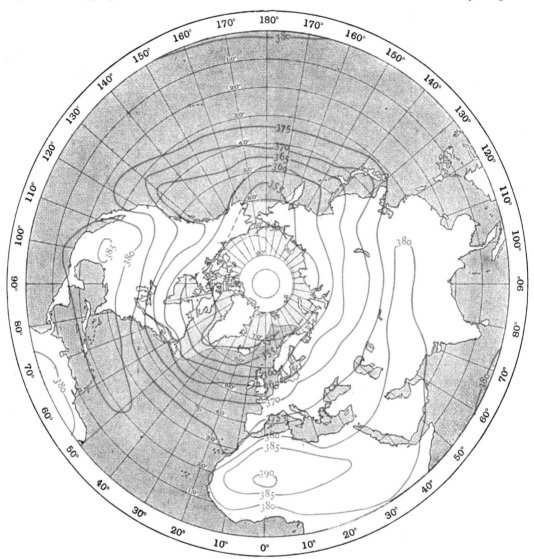

Equivalents			The pressure is computed from the pressure at the surface and the temperature with a lapse-rate of 5 a per 1000 m from 0 to 2000 m, 5·5 a from 2000 m to 4000 m, 6 a from 4000 m to 6000 m and 7 a from 6000 m to 8000 m. Allowance is made for the variation of gravity with latitude and with height but none for humidity.	Equivalents		
mb	mm	in		mb	mm	in
390	292·5	11·517		365	273·8	10·779
385	288·8	·369		360	270·0	·631
380	285·0	·222		355	266·3	·484
375	281·3	·074		350	262·5	·336
370	277·5	10·926		345	258·8	10·188

Fig. 61. Isobars for steps of five millibars at the level of 8000 metres in the northern hemisphere in July.

NORMAL PRESSURE OF THE LOWER LAYER: SEA-LEVEL TO 8000 METRES   PLATE XIX

Authority: Original.   *Reference to text*, pp. 78, 130.

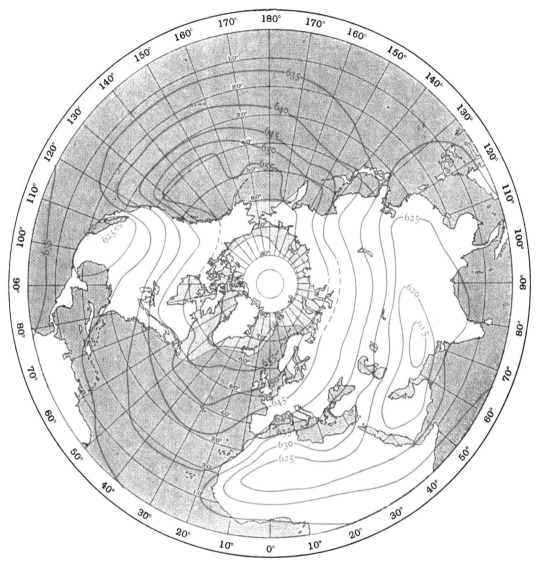

Note the analogy in the shape of the curves with those representing the pressure at 8000 metres, Fig. 61, Plate XVIII.

The loss of pressure between sea-level and 8000 metres is computed from the pressure at the surface (Fig. 29, Plate XXI), allowing for temperature with a lapse-rate, etc. as indicated on Fig. 61.

Fig. 62. Difference of pressure for 8000 metres from sea-level represented by isobars with steps of five millibars.

JANUARY

PLATE XX   NORMAL PRESSURE AT SEA-LEVEL IN THE NORTHERN HEMISPHERE

Reference to text, pp. 41, 71, 143.   *Authority:* M.O. compilation.

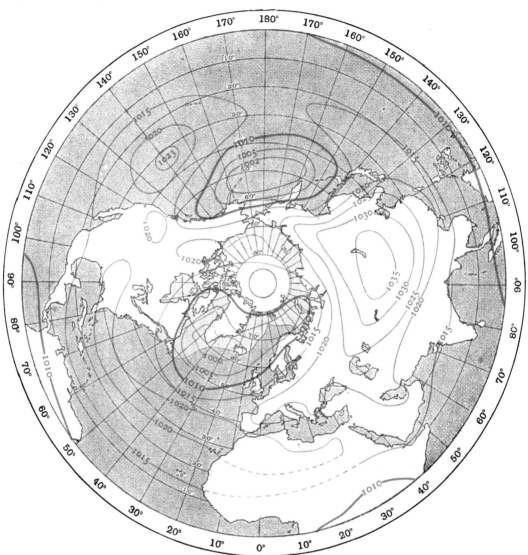

Equivalents		29·53 in = 1000 mb = 750 mm	Equivalents	
mb	in	The charts of normal pressure—Figs. 28 to 31—are based upon the charts of the *Barometer Manual for the Use of Seamen* (M.O. 61). Account has been taken of more recent information as indicated on p. 54.	mb	in
1005	29·68		1023	30·21
1010	29·83		1025	30·27
1015	29·97		1030	30·42
1020	30·12		1035	30·56

Fig. 28. Normal isobars in steps of five millibars for sea-level in the northern hemisphere in January.

JULY

NORMAL PRESSURE AT SEA-LEVEL IN THE NORTHERN HEMISPHERE   PLATE XXI

Authority: M.O. compilation.   *Reference to text,* pp. 41, 68, 71, 143.

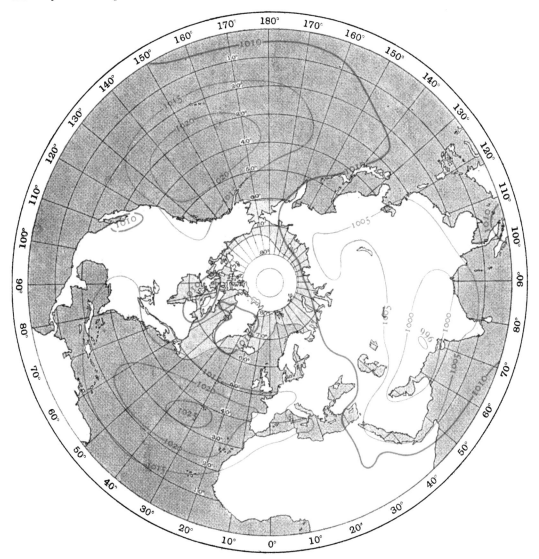

Equivalents		29·53 in = 1000 mb = 750 mm	Equivalents	
mb	in	The charts of normal pressure—Figs. 28 to 31—are based upon the charts of the *Barometer Manual for the Use of Seamen* (M.O. 61). Account has been taken of more recent information as indicated on p. 54.	mb	in
996	29·41		1015	29·97
1005	29·68		1020	30·12
1010	29·83		1025	30·27

Fig. 29. Normal isobars in steps of five millibars for sea-level in the northern hemisphere in July.

JANUARY

PLATE **XXII**   NORMAL PRESSURE AT SEA-LEVEL IN THE SOUTHERN HEMISPHERE

Reference to text, pp. 41, 143.   *Authority:* M.O. compilation.

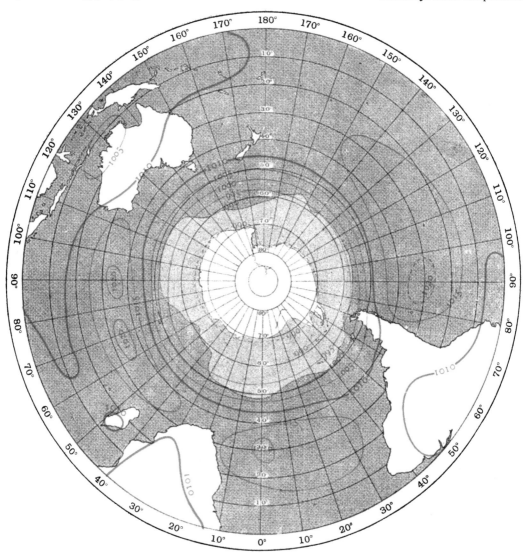

Equivalents		29·53 in = 1000 mb = 750 mm	Equivalents	
mb	in	The charts of normal pressure are based upon those given in the *Barometer Manual for the Use of Seamen* (M.O. 61). See Figs. 28 and 29.	mb	in
990	29·24		1010	29·83
995	29·38		1015	29·97
1005	29·68		1020	30·12

Fig. 30. Normal isobars in steps of five millibars for sea-level in the southern hemisphere in January.

NORMAL PRESSURE AT SEA-LEVEL IN THE SOUTHERN HEMISPHERE   PLATE **XXIII**

Authority: M.O. compilation.   *Reference to text,* pp. 41, 143.

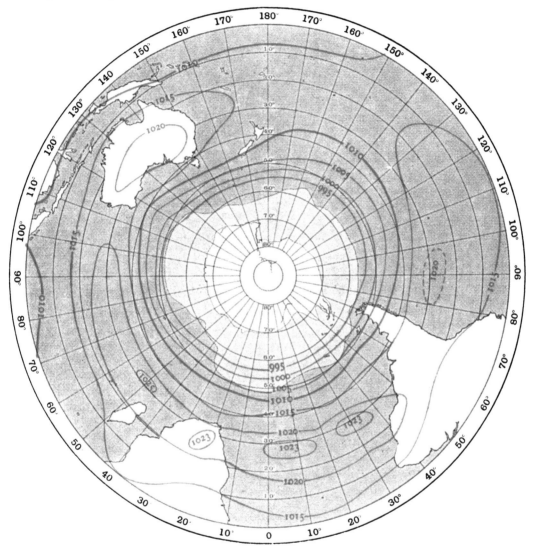

Equivalents		29·53 in = 1000 mb = 750 mm	Equivalents	
mb	in	The charts of normal pressure are based upon those	mb	in
995	29·38	given in the *Barometer Manual for the Use of Seamen*	1015	29·97
1000	29·53	(M.O. 61). See Figs. 28 and 29.	1020	30·12
1005	29·68		1023	30·21
1010	29·83		1025	30·27

Fig. 31. Normal isobars in steps of five millibars for sea-level in the southern hemisphere in July.

JULY

PLATE XXIV   NORMAL RESULTANT-FLOW OF AIR OVER THE NORTHERN HEMISPHERE

Reference to text, p. 68.   *Authority:* Original.

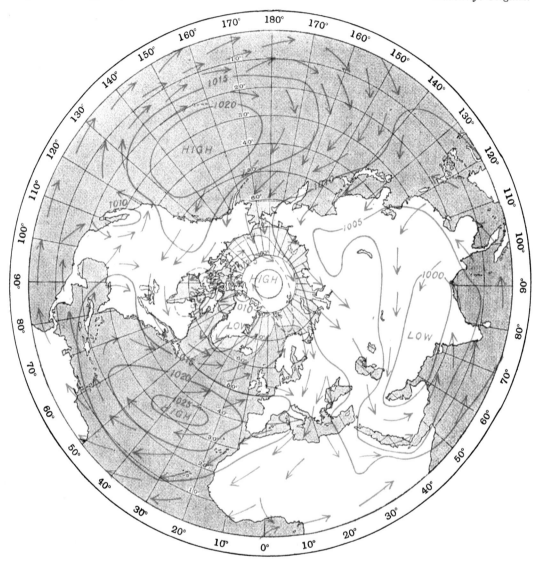

The arrows show the direction of the wind: they fly *with the wind* and are drawn according to the run of the isobars.

The isobars for this chart are taken from the lines of Plate XXI but the run of the line for 1010 mb in the equatorial region has been slightly modified to bring it into agreement with Fig 35, p. 54.

Fig. 43. Chart of the resultant-flow of air over the northern hemisphere in July according to the normal distribution of pressure.

1. METEOROLOGY FOR SCHOOLS AND COLLEGES

A PAPER CONTRIBUTED TO THE ROYAL METEOROLOGICAL SOCIETY IN RESPONSE
TO A REQUEST BY THE PRESIDENT. REPRINTED FROM THE *Quarterly Journal*
OF THE SOCIETY, VOL. XLIII, p. 83, 1917

WHENEVER the country decides that the Study of Weather is a subject of national importance, and judging by the experiences of the War the time is not far distant, it will be necessary for meteorologists to arrange the various divisions of the science in some sort of order according to the stage of mental development which the student may be assumed to have reached. Hitherto British books which have Meteorology for their title have been addressed to the general public and have avoided as far as possible any assumption of preliminary technical knowledge. The authors have had to explain physical processes and other things as they went along or to assume that no explanation is necessary. A common practice has been to lay down brief statements as a sort of concrete foundation of physical principle upon which to build a superstructure of explanation. When one looks into the matter all the processes of weather turn out to be much more complicated than the early meteorologists thought them to be. For example, when you have satisfied the inquiring mind, by some form of demonstration, that air becomes colder when it is rarefied dynamically it seems quite easy to extend the idea to explain that when a current of air flows over a mountain-chain it gets rarefied by elevation, and consequently cooled, with rainfall as the result. So certain are we of the soundness of the explanation that we have given the special name of orographic rainfall to the precipitation produced in that way. But when you come to think of it, the explanation requires that the air on the windward side has to be made to flow up-hill, and no fluid which technically must be called heavy, as it is affected by gravity, even if it is as light as air, flows up-hill without protest. It prefers to go round, and will exhaust all the possibilities of doing so before submitting to be driven over. As, however, the rainfall itself shows that air does get over mountain ranges, the possibilities are obviously exhausted, but the protest is somehow or other recorded; and before we regard the explanation as complete we ought to know what form the protest takes.

Or, to take another example: there are many meteorological processes which have been disposed of by the simple statement that warm air is lighter, bulk for bulk, than cold air, and moist air than dry air, and therefore warm air or moist air will rise. So it will in the comfortable environment of a physical lecture-room, but if you are applying the principle to explain the phenomena of weather you must be prepared for the inquiries how far will it rise, why does it ever stop, and why apparently is the warm air of the Sahara or the moist air of a London fog so reluctant to betake itself to the upper regions?

According to my recollection of physical geography at school we found no difficulty in accepting these summary explanations. True or untrue, adequate or inadequate, they helped us to remember the facts which were presented to us in maps which I still regard as the most enchanting form of literature. If I wanted to show benevolence to the British schoolboy I would take care that each one of him—or perhaps, better, every other one—should have an atlas of his very own when he is about ten years old, and as he grows up always a bigger and a better atlas, but not too big for him to carry easily. The glamour of the first possession of a physical atlas is still in my memory. It showed us oceans and their depths, continents and their heights, ocean currents and ocean winds, regions of perpetual snow, volcanoes, uninhabitable deserts, rivers and their courses, heights where trees grew or beasts wandered or birds soared, isotherms of mean temperature; pressure had not then got into maps. We asked for no references; we wanted no authorities; what was on the map might be explained, but it was not to be gainsaid or doubted. After years of implicit faith in them I wonder now to myself what an isotherm of mean temperature really means, who made it, and how did he draw those lines so firmly where the foot of civilised man had never trodden; who were the travellers who found out about the snow-lines and the trees and beasts and birds, and what is the authority for those ordered currents of wind and water? Fifty years ago, why should my enjoyment of the facts be marred by questionings about the authorities for them? That I should ever have lived to go behind the figures displayed in the physical atlas of my school days and even be the means of making better ones for the schoolboys who will come after me is a stupendous thought, as I look back.

I would still not seek to disturb a schoolboy's confidence in his maps. So long as you are dealing with the facts merely as facts, if the maps represent the best facts available we need not trouble ourselves, because, fifty years hence, some officious director of a meteorological office may draw the lines and arrows somewhat differently. If we like to draw isobars crossing a range of mountains without even a tiny waggle as an expression of protest I see no objection, provided we do not afterwards quote it as evidence that no protest is made in real life. And the like is true about climatic diagrams and climatological tables. For school purposes we need not wait until that far-off time when they will be quite beyond the reproach of the most competent expert. But things are quite different as soon as we begin to deal with the physical explanation of the atmospheric processes. It then becomes of the highest importance to examine and test every fact and figure. For example, in the climatological tables appended to *The Weather Map*[1], recently published, there are figures for absolute humidity which seem to show that, on the average, in our own climate the air gains moisture from daybreak onwards and loses moisture in the night hours, whereas the opposite is the case in the climate of Helwan in Lower Egypt. Now this may be, and very likely is, a real phenomenon that is susceptible of physical explanation; any one of us could, in

[1] M.O. Publication, No. 225. H.M. Stationery Office.

fact, offer a qualitative account of the process; but if we wished to verify it by actual figures it would be necessary to remember that the figures are obtained in each case from readings of the wet and dry bulbs, that they are reduced by different hygrometric tables, and that the dry climate of Helwan presents the best opportunity on earth for exhibiting the differences between the two sets of tables. So our attitude must be different according as we regard maps, diagrams, and tables merely as the representations of the best available facts or the final and actual basis of the physical explanation of the atmospheric circulation with all its incidental phenomena.

The physical explanations which were given in my school-days are, I believe, mostly fairy tales. I call them fairy tales because they deal so simply and swiftly with situations that in ordinary nature are dreadfully complicated. I see no harm in using fairy tales as a sort of connective tissue to help young boys to keep facts in their memories. Some sort of plausible story is necessary to satisfy the question Why? which is natural, not only to young boys, but to everybody who is not disciplined to be content with the answer to How? Of course, the tales ought to have as much verisimilitude as circumstances permit. If it could be avoided, I would not risk hurting Cinderella's pretty foot by a glass slipper that was obviously chipped. The worst of the fairy tales of physical geography was, not that schoolboys used them to string facts together with, but that they were accepted without verification by grown-up meteorologists as the basis of the dynamical explanation of the facts of the atmospheric circulation[1]. We have no use for fairy tales when dealing with fully reasoned physical explanations. The sovereign rule in that case is to prove all things, distrust everything that is not strictly proved, and if the particular question requires more accurate facts than those which are available wait till you have them, and meanwhile try something else, though it may be less ambitious.

The testing of generalisations and the development of theories may well be regarded as marking the difference between the meteorology of schools and the meteorology of colleges when the time comes. Yet I do not mean that it would be well to draw a hard and fast line. The boys of the upper classes of schools are sometimes equipped with adequate knowledge for the commencement of rigorous theory; they have the necessary knowledge of the laws and principles of dynamics and physics.

At the celebration of the fiftieth anniversary of the Society I was much struck by a remark of the President, the late Dr Theodore Williams. He commended meteorology because it was so easy; it could be pursued by an intelligent gardener. The remark is true enough for that part of meteorology which consists in the compilation of trustworthy observations on an organised

---

[1] For example, it is customary to regard the ascent of air and consequent rainfall of the doldrums of the Atlantic as one of the controlling factors of the general circulation, but I find Professor Cleveland Abbé, writing from personal experience (*Hann Band*, p. 258), says, "The isolated rainclouds from which showers fall in the doldrums do not necessarily represent important general ascending currents. They rarely occur at night-time, and are too infrequent and too small to represent any considerable part of the immense masses of air that flow toward the equator."

plan and their representation in tables, maps, and diagrams; that part, in fact, which I have called Meteorology for Schools. But the other part, the tracing of the physical causes of observed effects, which I have called Meteorology for Colleges, is of a different character. In the course of the last year I have given a general outline of what is understood by modern meteorology in *The Weather Map*, and have found it necessary, by way of explanation, to compile an auxiliary Glossary in which are collected some of the general principles and materials for the dynamical explanations. Any one who goes through that process will realise how easy and straightforward is the path, so long as we are dealing with the collection of facts, the formation of tables, maps, and diagrams—as far, in fact, as the compilation of a weather map and the lessons to be drawn directly from the study of maps, which include the principles of forecasting by means of weather charts. All these things are quite within the range of schoolboys, not of course to initiate, but to recapitulate; and they form already, as a matter of fact, a part of the curriculum in many schools. The maps, tables, and diagrams included in *The Weather Map* are a sufficient indication of that division of the subject. But let us go a step further and consider what the equipment of a student of the physical processes must be.

That part of physical geography which deals with the shape of the earth and her rotation, night and day, and the measurement of time, the motion of the earth in her orbit and the seasons, must be familiar to him. They will involve some knowledge of geometry and trigonometry. He must know something about the composition of the atmosphere, water-vapour, vapour-pressure, saturation, condensation, evaporation, and the numerical expression of these quantities, not at all an easy subject until familiarity has made it so.

But perhaps "absolute temperature" is the best text from which to start. So long as he deals only with observations, tables, maps, and diagrams, a meteorologist is apt to say, "Why worry me about absolute temperature? Why not let me measure temperature in the Fahrenheit scale which I understand?" To which the reply is another question: "How do you deal with the expansion of gases? Do you work out your questions of the expansion of gases on the Fahrenheit scale?—in schools they never teach that. How do you compute the density of air?" The whole of the circulation of the atmosphere depends upon the gaseous laws and upon variations of density, they appear in almost every meteorological calculation. You cannot take a single step in the explanation of the phenomena of weather without them; and if you wish to get beyond the stage of the schoolboy the gaseous laws must be constantly in your mind. You will require also to be familiar with the dynamical and thermodynamical properties of gases, and, still more recondite, the thermodynamical properties of moist air: it is useless to think about the formation of rainfall as a physical process without them. The great law of the conservation of energy must be your familiar friend, and absolute temperature then becomes a method of abbreviation.

There are other subjects which are just as valuable. One of the most important is an experimental knowledge of the mechanics of solids and fluids, fluid pressure, gravity and the motion of bodies under gravity in various circumstances; the motion of bodies under balanced forces, a section of dynamics which is very inadequately treated in text-books for schools and colleges. They are apt to confine themselves to motion in a straight line under no forces or under a constant force, because these lend themselves to easy computation; but the motion of the atmosphere never agrees with that limitation, and that form of simplification is impossible for a meteorologist. Then again the text-books are apt to leave out friction in order to make computation easy. Nature never leaves out friction and yet does the necessary computation without difficulty. If our friends in the colleges would only deal with experimental mechanics instead of limiting themselves to hypothetical mechanics, the comprehension of the phenomena of weather would be much easier. They probably forget that, like all other children, they began life with experimenting upon gravity in their cradles, to the worry of their nurses who had repeatedly to pick up things for them to drop once more.

There are some dynamical principles of universal application in meteorology which are only dealt with in colleges in the recondite region of rigid dynamics, but the results in nature are plain enough, and people who really understand them could, with the aid of practical illustration, bring their comprehension within the range of the meteorologist. Another great branch of dynamics in which we are interested is eddy-motion. We cannot pursue the study of meteorology very far if we ignore it. It is also very difficult, but even difficult subjects can be dealt with by apposite illustration.

Since Maxwell died, so far as I know, there has been nobody who has tried to put scientific reasoning in a form which could be comprehended by people unfamiliar with the forms of mathematics. By his *Theory of Heat* and *Matter and Motion* he opened a way which nobody follows, unfortunately for the amateur who wants to understand things.

There is the great subject of solar and terrestrial radiation. It is very difficult, but it is of such vital importance to us that, difficult or not, we must learn something of it.

Another great branch of the subject is meteorological optics—according to the *Scientific American* we are very ignorant about it in this country, and that may well be true, for the standard works on the subject are not English, and we have no one to make a text-book for us. It implies understanding something of refraction and diffraction and dispersion, all of which can be most beautifully illustrated by experiment. Since light consists of wave-motion, we might include with meteorological optics the phenomena of sound in the atmosphere, which also consists of wave-motion, the peculiarities of its transmission and their relation to the state of the weather.

And finally atmospheric electricity, ionisation, and lightning, the mysteries of the electric charges and discharges of the atmosphere with wireless telegraphy as a method of recording.

Thus, the equipment for which a fully armed student of atmospheric physics will find a use is rather formidable; it is beyond the ordinary resources of the gardener or the schoolboy, but it is not by any means necessary to have the whole complete before taking up some of the problems of the real atmosphere. Let us go back and consider how things might be arranged.

Meteorological optics with sound and atmospheric electricity may be regarded as separate sections that can be treated independently. So far as we know, the phenomena which are included in those sections are incidental to the general circulation; they do not transfer much of its energy, though the market value at current rates of the electrical energy of a single mile of lightning flash has recently been estimated by Mr R. A. W. Watt at £900, and there are many miles of lightning-discharges in the course of a year. For these sections all that we need is that enough people shall be interested in meteorology to make it worth while for some enterprising publisher to provide books on the subjects, specially written for meteorologists with the necessary physical introductions.

But we want at once a masterly chapter on radiation, because the whole of the circulation owes its motive power, though not the details of its form, to the radiation received from the sun. The facts and figures about the relation of radiation to the atmosphere are multitudinous. They are to be found in the literature of physics, and they have been co-ordinated, but not in a way that students of the atmosphere can readily follow.

The rest of the subjects mentioned, together with the observations, tables, maps, and diagrams of Meteorology for Schools, form the groundwork of the study of the physics of the atmosphere. Let me attempt briefly to indicate what a schoolboy should do to make it his own and qualify himself for studying college-meteorology.

First of all, he might get a slide-rule and become familiar by its use with what it all means. It includes trigonometry, the practical measurement of angles, and logarithms. Logarithms are particularly important, because the laws of nature have given the atmosphere a character which is peculiarly logarithmic. As one comes downward from external space to the solid earth by equal steps the pressure and density of the atmosphere progress by equal logarithmic increments, slightly modified by considerations of temperature and humidity. So logarithms are not merely a mathematical device for doing sums; they are a natural reality.

Then the laws of mechanics and of hydrostatics; the properties of motion of solids and fluids studied experimentally, including the phenomena of wave-motion and of eddy-motion, even if we cannot put them in the form of algebraical equations. These he will bear in mind while he is collecting and arranging a set of photographs of clouds of all sorts and varieties. And he will not omit to study the forces which are due to the flow of air or water past solid obstacles and the effect of air upon falling water-drops.

And, finally, the science of heat; thermodynamics and the laws of energy: that has been set out for us in a way that leaves little to be desired in Preston's

Theory of Heat. It needs an additional chapter on the thermodynamics of a mixture of air and water which has been set out by Hertz and subsequently by Neuhoff on the assumption that the reader is familiar with the manipulation of differential equations. But the differential equations only express in algebra facts of the kind with which, in reality, even the schoolboy is familiar in the form of graphs. Maxwell has had no successor in the endeavour to substitute diagrams for differential equations, because the teaching and learning of the subject have been practically confined to people who aspire to become physicists by profession, and for them it is a sign of weakness to miss an opportunity of using a differential equation; but when the nation shall arrive at the conclusion that Meteorology for Colleges is a subject which demands consideration it need not be doubted that a successor to Maxwell will be found.

Though the gap between the Meteorology of Schools and the Meteorology of Colleges is a wide one it is worth while to make an attempt to bridge it, because the explanation of natural phenomena is one of the irrepressible instincts of mankind. There is all the difference in the world between the physical explanation which gives a fully reasoned relation between cause and effect, which deals with measured quantities, and the imperfect explanations that may be fairy tales; and even if the old fairy tales should in the end prove substantially true there is much satisfaction to be derived from knowing exactly what they mean.

There is a large section of meteorology which is between that of schools and that of colleges. It is the section which deals with the arithmetical manipulation of accumulated facts, with the detection of periodic changes by the methods of harmonic analysis or otherwise, and the detection of the relation of different series of numbers by the methods of correlation. The actual processes involved are not beyond the capacity of a schoolboy, and the proofs of the formulae employed are within the reach of the highest classes in schools. But the selection of the materials for work of this kind requires judgment and experience. Any one who plunges into it without careful guidance is very apt to find that he has wasted time by proving something which was already known, or leaving things as vague as they were before. This kind of work is best undertaken by a student working under an experienced teacher of meteorology hardly to be found at present in colleges and not frequently in schools.

2. PRESSURE IN ABSOLUTE UNITS

REPRINTED FROM THE *Monthly Weather Review* OF THE U.S. WEATHER BUREAU
FOR JANUARY, 1914

FROM time to time, and especially within the last few years, the adoption of absolute units for representing atmospheric pressure has been urged on scientific grounds, and there is a general consensus of opinion that absolute units are the most suitable for dealing with meteorological theory, especially n relation to the upper air.

Through circumstances which are not altogether within my own control I have had to face the adoption of absolute units as a practical question and also as an educational question. In fact, I have had to ponder over replies to the following questions:

What units for pressure and temperature should be adopted in the publication of monthly values of pressure for a *réseau mondial?*

What units should be employed by lecturers and teachers who wish to interest students of mathematics and physics in the development of meteorological science?

What graduation should be employed for a barometer in order to commend most effectively to the wider public the results of meteorological study?

I find the answer to all these questions in absolute units on the C.G.S. system, with only an outstanding uncertainty as to whether the millibar or the centibar is to be preferred.

Perhaps I had better explain that the *bar* represents the C.G.S. "atmosphere," that is, a pressure of 1,000,000 dynes per square centimetre, the dyne being the C.G.S. unit of force. The dyne is the force which produces an acceleration of 1 centimetre per second per second, in a mass of 1 gram. The weight of *m* grams when the gravitational acceleration is *g* centimetres per second per second, is *mg* dynes. The bar is approximately equivalent to 750 millimetres, or 29·5 inches, of mercury at 0° C. and standard gravity. The centibar is one-hundredth, the millibar one-thousandth, of the bar.

It is quite possible that I may be to some extent affected by unconscious bias in favour of the ultimate application of theory to practice. If absolute units are the best for theory, they are the units of the future; for the practical applications of meteorology must ultimately be guided by theory, just as those of astronomy are at the present day. For me this supplies the answer to my first question. The time is coming, if it has not already come, when students of meteorology will deal with the earth as a whole on the basis of observations and will recognise that anything short of that is inadequate for the solution of the more general problems of climate and weather.

To my second question, as to what are the best units for educational purposes, there is only one answer. So far as the United Kingdom is concerned, in all schools and colleges, wherever the elements of mathematics, physics, and chemistry are instilled into the rising generation, they are in association with the metric system as a part of scientific education. Two consequences result therefrom: In the first place, a complete divorce of all scientific experience from the meteorological practice of everyday life, a divorce which may perhaps be sufficiently illustrated if I say that in the laboratory a water-bath of 98° is a very different thing from bath-water of 98° in everyday life. The whole of the disastrous effect of this divorce is hardly to be appreciated by those who have nearly accomplished their life's journey with comparative success in spite of that disadvantage, but that is no reason for disregarding its importance to the young, and therefore let me call special attention to another aspect of it.

Between professors and students of the mathematical and physical schools of our universities there is a "freemasonry," of which the use of metric units is a sign and from which the students of meteorology are apt to find themselves excluded. To express my meaning in the fewest words, let me say that if in a country assembly for the advancement of science, an unknown stranger should get up and speak in metric units, the initiated physicist would at once say "he must be one of *us*," and the uninitiated meteorologist would say "he is one of *them*"; but if he should begin his discourse by speaking in inches and grains, the physicists would at once say "we need not listen—there can be no dynamics or physics in this," and in the most out-of-the-way meteorological assembly, if any one should be heard speaking in metric units, he would not be set down as an eccentric or a crank, but as a person with exceptional scientific associations.

This being so, what should be the line of action of a meteorologist who lays claim to some portion of the scientific spirit? Surely this—not to remain in the isolation that excludes us from the sympathy of fellow-workers, but to turn the tables upon our friends and say to the grand masters of our cult, "We will accept a metric system, but we cannot accept your millimetre, because when we make a change we must take care not to perpetuate the unscientific practice of representing the pressure of the atmosphere by a length. We know that the millimetre which you use is not really a length at all, and is really only a millimetre under conventional conditions of temperature and latitude which never occur together, but our students, who have yet to learn that important fact, will have clearer ideas from the start if they do not begin with that confusion. We are prepared to do what physicists have often aspired to do, but have not had the courage or coherence to carry out, namely, to use pressure units for pressure measurements and leave length units to measure lengths with. Nor can we accept your centigrade scale with the freezing-point of water as its zero. We cannot let our students adopt the conception of negative temperatures, which is a survival of the time anterior to the conservation of energy and which has sooner or later to be explained away with much labour and practical inconvenience."

Let us now deal with the third question: What kind of barometer should be put before the general public with due regard for the teachings of modern meteorology? We know that it is still the practice to sell barometers with the customary legends:

28·0	28·5	29·0	29·5	30·0	30·5	31·0 inches
Stormy	Much rain	Rain	Change	Fair	Set fair	Very dry

and that many newspapers reproduce day by day a barometer dial of this kind. On metric barometers we find the same legends, but "Change" is opposite to 760 millimetres instead of 29·5 inches, and the steps are 10 millimetres instead of half inches. That is in itself sufficient condemnation of what on other grounds is quite intolerable, and in these days we want to suggest some alternative that will not spoil the instrument-makers' trade, nor yet convey to the countryman misleading ideas.

The first idea that an official meteorologist would suggest is that no countryman would have done his duty by the atmosphere unless he had compared his local reading with that of the corresponding issue of the daily bulletin. To do that he must reduce his readings to sea-level, so, absolutely, the first requirement is a simple means for giving, with sufficient approximation, the sea-level pressure. The next idea to be inculcated is that the actual pressure of the atmosphere at the moment does not matter as a general rule, but only the changes which are taking place, and which can be watched locally with great advantage. What could be better for this purpose than to mark some point within the range of the barometer 100 and note the differences from that point as percentages? Coming to details, it can only be regarded as providential that the point on the barometer against which the word "Change" is inscribed, being 29·5 inches, corresponds almost exactly with 100 centibars; consequently the temptation to use centibars and write 100 there is irresistible. Then obviously we must make the range of the dial or the tube big enough to show the changes which are to be expected in the district in which it is to be used, and the countryman will at once realise within what percentages of the middle value the pressure has varied in the past, and therefore may be expected to vary in the future. It is curious that 100 centibars, although not the mean value of the sea-level pressure, is in the middle of the usual range, and is, in fact, the middle line of the ordinary record-sheet of a Richard-barograph, which is marked 75 centimetres or 29·5 inches.

By way of suggesting that it is variations of the barometric pressure which count, and not the particular level, we can give the frequencies of occurrence of different barometric pressures, so that the observer can see for himself whether conditions are normal or exceptional, and so keep an eye on the working of his instrument as well as on the weather.

I have set out these suggestions in a *Land Barometer* (Fig. 1), with a rotating circle for reduction to sea-level. It is not necessary to enter into any further explanation; what is set out on the dial ought to be self-explanatory. But I ought to say a word about the frequencies. I cannot now recall where I got the figures which are engraved on the first dial. I have made new figures for subsequent specimens, which give the average frequency of barometric minima below 100 centibars and maxima above 102 centibars, for Valencia, Aberdeen, and Kew combined. The figures are not applicable to any particular place without further inquiry, and their entry on the dial is a challenge to the observer to verify or improve them for his own locality.

Fig. 1. Land Barometer.

The barometer is compensated for temperature and its readings are independent of latitude. It is therefore graduated in centibars and millibars. Scales of mercury inches and millimetres at 0° C. for latitude 45° are added for comparison. A sliding rim which is also graduated in centibars and millibars is provided in order that a reduction to sea-level may be made mechanically. In the spaces between the figures of the centibar graduation are given the frequencies of occurrence of sea-level pressures for those limits derived from observations at Kew, Valencia and Aberdeen taken together.

With the *Sea Barometer* (Fig. 2) things are different. The observer has no daily weather chart at hand to show the distribution of pressure at sea-level with which he can check the readings of his own barometer, and for any check he must rely upon the normals for his locality. Nor can frequencies of barometer values be easily given for a sailor whose course runs north or south. A barometer, on the pattern of the land barometer, for the transatlantic voyage might be made, but for the sailor who is not restricted to the transatlantic passage the normals for different latitudes along the thirtieth west meridian seem the most effective, and they are shown on the sea barometer dial.

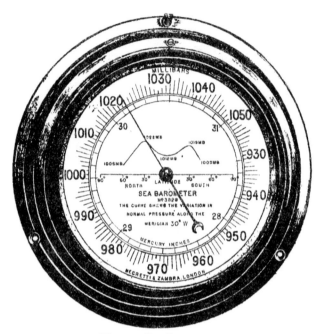

Fig. 2. Sea Barometer.

The barometer is compensated for temperature and its readings are independent of latitude. It is therefore graduated in millibars. A scale of mercury inches for latitude 45° is added for comparison.

On the central part of the dial is a curve which shows the variation of the mean annual pressures on the meridian of 30° W.

N.B. In these instruments the central figure 1000 is placed where it should be so that when the pressure is rising the pointer is going up and *vice versâ*, not at the top from which it goes down either way.

These two barometer dials seem to suggest the centibar as the c.g.s. unit most likely to be useful in practice. So far no one has had any effective experience with instruments graduated to give pressure in absolute measure, and the millibar has given its name to the battlefield between the old and new, because it was adopted by Prof. Bjerknes as a substitute for the millimetre. Fortunately the difference between centibars and millibars is only the difference of a decimal point, and the practice as to observing and publishing may be allowed to shape itself as convenience in practice dictates.

While I am writing on the subject of absolute units, I should like to add a word about the proposal of Prof. Bjerknes to record heights in "dynamic metres," which has given rise to fierce controversy. The quantity which it is really sought to express by the use of the term "dynamic metres" is the product gh, which is in fact the potential energy of unit mass at the height h. This quantity may quite appropriately be called the *geo-potential*, that is, the potential due to the earth's gravitational attraction at the height h. In the units which Prof. Bjerknes employs, the acceleration g becomes numerically 0·981 for latitude 45°, and if h is expressed in metres, gh differs from h by less than 2 per cent.

Thus the expression for the height in metres is numerically little different from the expression of the *geo-potential* in what Prof. Bjerknes calls *dynamic metres*. The objection to the suggestion may be briefly expressed by saying that what is sought to be represented is not really height as generally understood, for example, in a pilot balloon sounding, which is essentially a geometrical measurement, and the unit in which the geo-potential is expressed is not a *metre*, nor any fixed length; it has not the unitary "dimension" of a length.

The objection to changing the "dimension" of a unit by prefixing an adjective is perfectly sound, but it is really a curiosity of scientific literature to find the objection to the use of "dynamic metre" for the expression of geo-potential denounced as immoral in an article which stoutly upholds the use of the "time-honoured millimetre" as a unit of pressure, without even an adjective as a warning to the unwary.

Mr F. J. W. Whipple, of my office, has proposed a solution of the difficulty which seems to me to meet the case in a satisfactory way. He points out that we have no special name for the unit of acceleration, and that in quoting the acceleration of gravity for a particular latitude we have to express it as, say, 981 centimetres per second, per second. He suggests the name *leo*, an abbreviation of the name of Galileo, of immortal memory in connection with gravitation, as a suitable name for the acceleration of a dekametre per second, and in this unit the acceleration of gravity in latitude 45° would be 0·981 *leo*. Then on the analogy of the *kilogram-metre* or the *foot-pound*, both time-honoured as units of work or potential energy, a *leo-metre* would be the potential energy of unit mass raised through 1 metre against an acceleration of 1 "leo" or of unit mass raised through $1/g$ metres against the acceleration of gravity. Thus Prof. Bjerknes' "height" in *dynamic metres* would become the *geo-potential* in "*leo-metres*," and would differ numerically from the real height in metres only by about 2 per cent.

In this way all the objections on the score of morality or unsound terminology would be avoided, and yet the numerical value of the geo-potential in "leo-metres" would enable us to keep in mind a close approximation to the actual height in the consideration of the dynamic problems of the atmosphere.

It seems clear that the time has come when meteorologists may properly turn their attention to the reconsideration of their units and their nomenclature, and that the call comes with almost equal force from the theoretical, the educational, and the practical sides of their work.

CHANGE OF UNITS OF MEASUREMENT. MAY, 1914.
M.O. Circular 202

Barometric Pressure in Pressure Units

In their Eighth Report to the Lords Commissioners of His Majesty's Treasury, the Meteorological Committee intimated their intention to use Absolute Units for pressure in the Daily Weather Report of the Meteorological Office from 1st May, 1914.

The absolute unit of pressure on the Centimetre-Gramme-Second system[1] is the dyne per square centimetre. As this unit is exceedingly small a practical unit one million times as great has been suggested. This unit, the megadyne per square centimetre, is called a "bar." In the Daily Weather Report the centibar and the millibar, respectively, the hundredth and the thousandth part of the "bar," are adopted as working units. The relation between the millibar and the inch of mercury is given in the tables overleaf.

Reasons for the Change

One of the principal reasons for this change is that it is a step towards the adoption of a system of units which may become common to all nations.

The system was approved by the Meteorological Council in 1904 and by the Gassiot Committee of the Royal Society in 1910. Upon the initiative of Prof. V. Bjerknes, formerly professor at Christiania, and now of the Geophysical Institute at Leipzig, it was used in important publications of the Carnegie Institution of Washington, and was adopted by the International Commission for Scientific Aeronautics for the international publication of the results of the investigation of the upper air. Since 1907 the system has been used in the Meteorological Office for the upper air, and since 1911 for the data from the observatories where c.g.s. units have been used for many years in connection with magnetism and electricity. The Weather Bureau of the United States has adopted *millibars* and *absolute temperatures* on the centigrade scale for the issue of daily charts of the Northern Hemisphere, which began on 1st January, 1914; the Royal Meteorological Society has decided to use *millibars* for the expression of the series of pressure normals for the British Isles, which it is now preparing; and the Meteorological Office has followed the example of the Weather Bureau in using absolute units for the daily maps in the Weekly Weather Report, but its isobars are figured in centibars as they were in the specimen issued with the Eighth Annual Report.

The Scientific Appeal

The ground of scientific appeal to all nations to adopt the bar, centibar, and millibar is that these units fall naturally into place as members of the c.g.s. system of units which has already become universal for magnetism and electricity and most branches of physics. Its principles are therefore well known. The inch and the millimetre are really units of length, and to estimate

[1] Particulars of the Centimetre-Gramme-Second system are given in the *Observer's Handbook*, 1913 edition.

the effect of a pressure measured in terms of height of a column of mercury, it is necessary to introduce the value of the density of mercury at some particular temperature and the value of the acceleration due to gravity at a particular place. It is well known that the atmospheric pressure at sea-level in Britain varies between 13¾ and 15¼ lbs. weight per square inch. The pound weight per square inch is often used by engineers, but it is not a convenient unit because its value depends upon latitude.

The Upper Air

The past fifteen years have witnessed the collection of extensive meteorological observations in the upper air, made by means of kites and balloons, from which important results have already been deduced. The absolute system of units is the most convenient for the discussion of the data so collected, and it is being generally adopted for the purpose. The rapid development of aviation makes it impossible to draw a line between the academic study of the meteorology of the upper air and the practical meteorology of the Daily Weather Report. The use of two systems of units, one for observations made at the surface, and the other for observations taken at higher levels, could only retard progress.

Practical Considerations

It is acknowledged that an accuracy of one-thousandth of an inch is not really attainable in practice. For many years the Inspectors of the Meteorological Office have had to be satisfied with agreement within ·003 in., and now the National Physical Laboratory has ceased to certify barometers of the Kew pattern to the thousandth of an inch. Consequently, with an instrument graduated to ·001 in., observers are being asked to read to an accuracy which is acknowledged to be unattainable. On the other hand an accuracy of the hundredth of an inch is not good enough for scientific purposes.

The practical degree of precision for a mercury barometer of the Kew type is *one-tenth of a millibar*. Graduation in centibars and millibars, with a simple vernier scale for estimating to tenths of a millibar, thus brings the demand for accuracy made upon the observer into harmony with that actually attainable. The new graduation does away with the complications of the conventional vernier scale in use on barometers graduated in inches, and consequently the risk of errors of observation is reduced.

The Percentage Barometer

Another advantage is that the bar, or c.g.s. atmosphere, differs but little from the standard atmosphere. The equivalent of the adopted normal value at sea-level of 29·92 mercury inches is 101·32 centibars, or 1013·2 millibars. The lowest barometer value ever observed for sea-level in the British Isles is 925·5 millibars, the equivalent of 27·33 inches. This value was recorded at Ochtertyre on 26th January, 1884. The highest value is 1053·5 millibars, the equivalent of 31·11 mercury inches. It was recorded at Aberdeen on 31st January, 1902.

A reading of 100 centibars, or 1000 millibars, is equivalent to 29·53 mercury inches. It will be remembered that the word "Change" is placed opposite the sea-level reading 29·5 in the conventional descriptions engraved on dial barometers. Thus in a barometer graduated in centibars the reading 100 would occupy the position conventionally marked "Change."

Practical Course to be pursued

It is evidently impossible at one operation to change all the barometers in use in the various services, and even in the most favourable circumstances there must be for many observers a time when the readings are taken on one scale, and the results quoted or published in another. Tables of equivalents are given herewith for making the necessary conversion.

The barometers issued by the Meteorological Office will be graduated in both scales[1].

RAINFALL DATA IN MILLIMETRES

As a further step in the direction of international uniformity all rainfall data will be published in the Daily Weather Report in millimetres instead of inches. The occasion for making the change is that modifications are being introduced into the telegraphic code used for the exchange of meteorological information in Europe.

The reading of rainfall in this country has been carried to hundredths, sometimes to thousandths of an inch, but the readings to the higher degree of accuracy have seldom any practical meaning. The readings on the metric system are carried to 0·1 millimetre 0·004 inch, which represents satisfactorily the highest degree of accuracy. The range is from ·01 to 3, 4, or even more inches in exceptional circumstances, for a day's rain. The telegraphic code hitherto in use has made provision for reporting amounts up to 10 inches, though the large majority of the readings are under 2 inches. The code now to be introduced makes provision for reporting amounts up to 100 millimetres or 4 inches

As one inch is approximately equivalent to 25 millimetres the conversion from millimetres to inches, or *vice versâ*, may be made with sufficient accuracy for most purposes by multiplying or dividing by four and appropriately shifting the decimal point. Tables of conversion are given herewith.

WIND VELOCITIES IN METRES PER SECOND

Wind force will be specified on the Beaufort scale. Occasional reports are received from anemometer stations regarding the extreme wind velocities attained in gales. These data are published on the front page of the report. The unit of wind velocity used in such cases will be the *metre per second*. Tables for converting velocities from miles per hour to metres per second, or *vice versâ*, are given below.

[1] The inch scale, in accordance with universal custom, is made to be *correct* when the brass case is at 62° F. and the mercury at 32° F. The millibar scale was made at first to be correct when both scale and mercury were at 273 a. In either case a large correction was usually necessary. It soon became apparent that this was not only cumbrous but quite unnecessary and now the millibar scale is graduated to be correct at the ordinary temperature at which it is read. This temperature is the "fiducial temperature" for the particular barometer and as working temperatures differ little from the fiducial temperature the corrections are usually small.

Pressure Values
Equivalents in Mercury Inches at 32° F. and Latitude 45° of Millibars

Milli-bars	0	1	2	3	4	5	6	7	8	9
	Mercury inches									
910	26·87	26·90	26·93	26·96	26·99	27·02	27·05	27·08	27·11	27·14
920	27·17	27·20	27·23	27·26	27·29	27·32	27·35	27·38	27·41	27·44
930	27·46	27·49	27·52	27·55	27·58	27·61	27·64	27·67	27·70	27·73
940	27·76	27·79	27·82	27·85	27·88	27·91	27·94	27·97	28·00	28·03
950	28·05	28·08	28·11	28·14	28·17	28·20	28·23	28·26	28·29	28·32
960	28·35	28·38	28·41	28·44	28·47	28·50	28·53	28·56	28·59	28·62
970	28·65	28·67	28·70	28·73	28·76	28·79	28·82	28·85	28·88	28·91
980	28·94	28·97	29·00	29·03	29·06	29·09	29·12	29·15	29·18	29·21
990	29·24	29·26	29·29	29·32	29·35	29·38	29·41	29·44	29·47	29·50
1000	29·53	29·56	29·59	29·62	29·65	29·68	29·71	29·74	29·77	29·80
1010	29·83	29·86	29·89	29·92	29·94	29·97	30·00	30·03	30·06	30·09
1020	30·12	30·15	30·18	30·21	30·24	30·27	30·30	30·33	30·36	30·39
1030	30·42	30·45	30·48	30·51	30·53	30·56	30·59	30·62	30·65	30·68
1040	30·71	30·74	30·77	30·80	30·83	30·86	30·89	30·92	30·95	30·98
1050	31·01	31·04	31·07	31·10	31·13	31·16	31·18	31·21	31·24	31·27

Differences for tenths of a millibar:

Mb.	·1	·2	·3	·4	·5	·6	·7	·8	·9
In.	·003	·006	·009	·012	·015	·018	·021	·024	·027

Rainfall Values
Equivalents in Inches of Millimetres

Milli-metres	0	1	2	3	4	5	6	7	8	9
	Inches									
0	0·00	0·04	0·08	0·12	0·16	0·20	0·24	0·28	0·32	0·35
10	0·39	0·43	0·47	0·51	0·55	0·59	0·63	0·67	0·71	0·75
20	0·79	0·83	0·87	0·91	0·95	0·98	1·02	1·06	1·10	1·14
30	1·18	1·22	1·26	1·30	1·34	1·38	1·42	1·46	1·50	1·54
40	1·58	1·61	1·65	1·69	1·73	1·77	1·81	1·85	1·89	1·93
50	1·97	2·01	2·05	2·09	2·13	2·17	2·21	2·24	2·28	2·32
60	2·36	2·40	2·44	2·48	2·52	2·56	2·60	2·64	2·68	2·72
70	2·76	2·80	2·84	2·87	2·91	2·95	2·99	3·03	3·07	3·11
80	3·15	3·19	3·23	3·27	3·31	3·35	3·39	3·43	3·47	3·50
90	3·54	3·58	3·62	3·66	3·70	3·74	3·78	3·82	3·86	3·90

Wind Velocity
Equivalents of Metres-per-Second in Miles-per-Hour

Metres per second	0	1	2	3	4	5	6	7	8	9
	Miles per hour									
0	0·0	2·2	4·5	6·7	9·0	11·2	13·4	15·7	17·9	20·1
10	22·4	24·6	26·8	29·1	31·3	33·6	35·8	38·0	40·3	42·5
20	44·7	47·0	49·2	51·5	53·7	55·9	58·2	60·4	62·6	64·9
30	67·1	69·4	71·6	73·8	76·1	78·3	80·5	82·8	85·0	87·2
40	89·5	91·7	94·0	96·2	98·4	100·7	102·9	105·1	107·4	109·6

Specification of the Beaufort Scale of Wind Force with Probable Equivalents of the Numbers of the Scale

Beaufort number	General description of wind	Specification of Beaufort scale — For coast use, based on observations made at Scilly, Yarmouth and Holyhead	Specification of Beaufort scale — For use on land, based on observations made at land stations	Mean wind force at standard density — Mb.	Mean wind force at standard density — Lbs. per sq. ft.	Equivalent velocity in miles per hour	Limits of velocities — Statute miles per hour	Limits of velocities — Metres per second	Beaufort number
0	Calm ...	Calm ...	Calm; smoke rises vertically	0	0	0	Less than 1	Less than 0·3	0
1	Light air	Fishing smack* just has steerage-way	Direction of wind shown by smoke drift, but not by wind vanes	·01	·01	2	1-3	0·3-1·5	1
2	Slight breeze	Wind fills the sails of smacks, which then move at about 1-2 miles per hour	Wind felt on face; leaves rustle; ordinary vane moved by wind	·04	·08	5	4-7	1·6-3·3	2
3	Gentle breeze	Smacks begin to careen, and travel about 3-4 miles per hour	Leaves and small twigs in constant motion; wind extends light flag	·13	·28	10	8-12	3·4-5·4	3
4	Moderate breeze	Good working breeze; smacks carry all canvas, with good list	Raises dust and loose paper; small branches are moved	·32	·67	15	13-18	5·5-8·0	4
5	Fresh breeze	Smacks shorten sail ...	Small trees in leaf begin to sway; crested wavelets form on inland waters	·62	1·31	21	19-24	8·1-10·7	5
6	Strong breeze	Smacks have double reef in main sail. Care required when fishing	Large branches in motion; whistling heard in telegraph wires; umbrellas used with difficulty	1·1	2·3	27	25-31	10·8-13·8	6
7	High wind	Smacks remain in harbour, and those at sea lie to	Whole trees in motion; inconvenience felt when walking against wind	1·7	3·6	35	32-38	13·9-17·1	7
8	Gale	All smacks make for harbour, if near	Breaks twigs off trees; generally impedes progress	2·6	5·4	42	39-46	17·2-20·7	8
9	Strong gale	...	Slight structural damage occurs (chimney pots and slates removed)	3·7	7·7	50	47-54	20·8-24·4	9
10	Whole gale	...	Seldom experienced inland; trees uprooted; considerable structural damage occurs	5·0	10·5	59	55-63	24·5-28·4	10
11	Storm	...	Very rarely experienced; accompanied by widespread damage	6·7	14·0	68	64-75	28·5-33·5	11
12	Hurricane	...	...	8·1	Above 17·0	Above 75	Above 75	33·6 and above	12

* The fishing smack in this column may be taken as representing a trawler of average type and trim. For larger or smaller boats and for special circumstances allowance must be made. † For converting estimates on the Beaufort scale into miles per hour (anemometer factor, 2·2).
‡ For finding the Beaufort number corresponding to a velocity expressed in miles per hour.

3. THE METEOROLOGY OF THE GLOBE IN 1911

REPRINTED FROM THE *Quarterly Journal of the Royal Meteorological Society*,
VOL. XLII, No. 179, JULY 1916

THE title which has been chosen for me to describe the subject of this lecture is rather more ambitious than I am myself; the object which I had in view, in undertaking the responsible duty of lecturer at this exceptional time, was to make known to the society, and thereby to all those who are interested in meteorology, the completion of the first instalment of a work which has engaged the attention of the Meteorological Office for many years, and which is, in brief, the presentation of a collection of facts regarding pressure, temperature and rainfall to form the basis of a comprehensive study of the meteorology of the globe.

The year 1911 is still remembered in the British Isles and also in the United States for its fine, warm summer. A correspondence upon the subject was opened in the columns of *Nature* by Sir Edward Fry with a challenge to meteorologists to produce the reason for it. Then, as on so many previous occasions, the discussion meandered for a while and lost itself in the desert of want of facts about the weather in different parts of the world, and attention was thereby called once more to the necessity of a comprehensive and homogeneous representation of the meteorology of the globe so far as that is possible with the data which are provided by existing stations. The volume which provides the representation for the year 1911 is now completed and awaits only the printing of some illustrative charts[1]. It bears the title *Réseau Mondial*, which will recall to meteorologists the initiative of the lamented M. Teisserenc de Bort and of his colleague Professor Hildebrandsson in studies of this character.

The production of a set of comprehensive data for the meteorology of the globe is by no means a new object of ambition among meteorologists. It can be found underlying the deliberations of the International Conference at Leipzig in 1872 and the formal Congresses at Vienna in 1873 and Rome in 1879. These Congresses laid down the lines of organisation of meteorological stations, so that all countries which maintain meteorological observations, primarily for their own economic purpose, should co-operate in making public a selection of their data in common form. The contributions of the several countries to this object are represented by the publications of Observations at Stations of the Second Order, and, if the aspirations of the Congresses were fully achieved, a collection of the whole series for any year would afford material for a representation of the meteorology of the globe for that year which any one could arrange for his own purposes in his own way.

Various circumstances have combined to interfere with the realisation of

[1] *Réseau Mondial*—monthly and annual summaries of pressure, temperature, and precipitation at land stations, generally two for each ten-degree square of latitude and longitude. M.O. No. 207 g, 1916.

the object in that way. First, the publications provide very unequal numbers of stations of the second order[1] for different parts of the globe. Europe and certain parts of North America and Asia have an abundance of stations, other parts of the world hardly any; and the publication takes place irregularly. Some countries produce their data with exemplary promptitude, others only after several years' delay, so that the investigator who wishes to compile a representation of the meteorology of the globe needs to have, not a single volume but a whole shelf of volumes for a single year, and he would have to have a set of shelves before a single year was completed. Very few libraries in any country would be able to provide them, and even then the results would be incomplete because there are many countries which are of great importance in meteorological study, but are not very wealthy and so have no regular meteorological organisation and do not share in the international co-operation. Among these must be set a large number of our own smaller colonies.

The second circumstance which interferes with the realisation of the object is the difference of units. One half of the world uses different units for pressure, temperature and rainfall from the other half, so that any one who wishes to take out the data must convert one half of them before he can compare the results.

Thirdly, all meteorological data from individual stations require a critical examination to make sure of their being homogeneous with their neighbours. Observations of pressure may or may not be corrected for temperature and latitude and reduced to sea-level. The mean temperature is derivable from observations at fixed hours in a great variety of ways. Unless the results are critically examined within a short time of their being taken a number of pitfalls for the unwary student of the meteorology of the globe may be hidden in the figures.

One example will suffice. The data for a certain station in West Africa were published officially year by year, and in one year an error crept into the statement of the height of the station above sea-level by the transposition of the figures in printing. Then the position of the station was changed and a new height of the barometer was computed by adding the difference to the misprinted figures instead of the correct figures. In consequence all the subsequent values were affected by what is known technically as a systematic error very difficult to detect by any one who has not the whole series of values before him.

Much, therefore, was required in addition to the recommendations of the International Congresses before the representation of the meteorology of the globe for a single month or year could be provided for, and the necessity for the additional effort has been urged on many sides.

Among the most urgent appellants for further effort in this direction were Professor Hildebrandsson of Upsala and M. Teisserenc de Bort of Paris, who

---

[1] A station of the second order or normal climatological station is one at which observations of pressure, temperature, wind and weather are made twice or thrice daily at fixed hours through the year. The word "second" is used to distinguish it from a station of the first order at which continuous records are obtained or hourly observations made.

repeatedly called attention to the meteorological importance of certain regions of the globe as centres of action of the atmosphere. Professor v. Hann of Vienna lost no opportunity of adding to the material which is regularly published whatever could be gathered from out-of-the-way stations. In 1903 Sir Norman Lockyer called attention to the necessity of dealing with the meteorology of the globe as a whole in the consideration of questions of the relation of solar and terrestrial changes, and a Commission called the Solar Commission was appointed by the International Meteorological Committee to organise a special compilation of meteorological data for the globe as well as data relating to terrestrial magnetism and the sun.

In connection with the meetings of this Commission a scheme of representation of the meteorology of the globe on the basis of two stations for each square of ten degrees of latitude and longitude was elaborated and a selection of stations was drawn up. It fell to my lot to present a specimen table of data drawn up on this basis to the meeting of the International Meteorological Committee at Berlin in 1910, and it became my duty as President of the Committee to bring it into co-operation with the proposals for the collection of data for a Réseau Mondial made by Teisserenc de Bort for telegraphic reports and subsequently extended by Hildebrandsson to take in climatological data.

The specimen table for one month presented to the meeting of the International Meteorological Committee at Berlin in 1910 gave data for January 1905, and shortly after that meeting the question of the removal of the Solar Physics Observatory from South Kensington to Cambridge involved the readjustment of the responsibility of the Meteorological Office for the compilation and publication of the data. Some years elapsed before provision could be made for the subject to be dealt with systematically, and by that time the year 1905 had receded into a somewhat remote antiquity. The year 1911 forms an appropriate epoch for commencement because it is the beginning of a new decade, and sufficiently remote for the great majority of the official publications containing the data to have been issued. It is accordingly with 1911 that a beginning has been made.

The work has been in the charge of Mr C. E. P. Brooks, Librarian of the Meteorological Office, who has also taken charge of the returns, whether printed or in manuscript, from our own colonial stations.

The List of Stations

It was arranged by the Solar Commission that the meteorology of the globe should be represented on the basis of two stations for each ten-degree square of latitude and longitude, and a selection of stations was made. That selection has been adhered to in the present publication so far as adherence was practicable. A few stations were found not to be available, and in consequence others have been selected in their places. Out of a possible number of 1152 to represent the whole globe, 392 are available. The material now presented may be supplemented by observations at sea. Dr J. P. van der Stok of the Meteorological Institute of the Netherlands has taken that part of the work in hand,

and some statistics have already been published[1]. The effort deserves all possible encouragement.

Many of the squares have no land on which a permanent station can be maintained. The list of land stations is a very searching exercise in geography. The names in each zone read from 180° W. to 180° E. In spelling, the practice of the original authorities has been followed.

In Zone 1, latitude 70° to 80° N., there are 6 stations, viz.: North Star Bay, Upernivik, Spitsbergen, Gjesvaer, Mehavn, Vardö.

In Zone 2, latitude 60° to 70° N., there are 36 stations, viz.: Dawson, Carcross, Fort Simpson, Hay River, Jacobshavn, Godthaab, Angmagsalik, Stykkisholm, Vestmanno, Grimsey (Akureyri), Berufjord, Thorshavn, Bergen (Met. Obs.), Christiansund, Trondhjem, Bodö, Haparanda, Helsingfors, Kuopio, Kola, Petrozavodsk, Arkhangelsk, Velsk, Oust-Tsylma, Troitsko-Petcherskoe, Berezov, Obdorsk, Sourgout, Touroukhansk, Doudinka, Olekminsk, Iakoutsk, Verkhoïansk, Oust-Maïskoe, Markovo sur Anadyr, Novo-Mariinskii Post.

In Zone 3, latitude 50° to 60° N., there are 52 stations, viz.: Prince Rupert, Barkerville, Kamloops, Calgary, Fort Chipewyan, Prince Albert, Qu'Appelle, Minnedosa, Berens River, Fort Churchill, Fort Hope, Moose Factory, Hebron, Nain, Point Riche, Valencia, Aberdeen, Greenwich, Uccle, De Bilt, Hamburg, Copenhagen, Potsdam, Upsala, Warsaw, Jurjev, Petrograd, Kiev, Moscow, Saratov, Kazan, Orenburg, Perm, Ekaterinburg, Tobolsk, Akmolinsk, Omsk, Barnaoul, Tomsk, Minousinsk, Eniseisk, Irkutsk, Kirensk, Tchita, Blagovechtchensk Priisk, Nertchinsk, Tygan Ourkan, Blagovechtchensk, Paikanskii Sklad, Nikolaevsk-sur-Amour, Okhotsk, Petropavlovsk Phare.

In Zone 4, latitude 40° to 50° N., there are 43 stations, viz.: Victoria, B.C.; Portland, Oregon; Macleod, Helena, Salt Lake City, Cheyenne, North Platte, Bismarck, Winnipeg, Duluth, Port Arthur, Chicago, Toronto, New York, St John, N.B.; S.W. Point, Anticosti; Sable Island (Main Station), St Johns, Newfoundland; Madrid, Nantes, Paris (Parc St Maur), Marseilles, Zürich, Rome, Vienna, Buda-Pesth, Bucharest, Odessa (University), Kharkov (University), Novorossiisk, Tiflis, Astrakhan, Fort Alexandrovsk, Krasnovodsk, Tachkent, Irgiz, Narynskoe, Vernyi, Mukden, Joshin, Vladivostok, Ochiai, Nemuro.

In Zone 5, latitude 30° to 40° N., there are 46 stations, viz.: San Francisco, San Luis Obispo, Modena, San Diego, Santa Fé, Denver, Abilene, Saint Louis, Mobile, Nashville, Charleston, Washington, Bermuda, Horta, Ponta Delgada, Madeira (Funchal), Lisbon, Casablanca, Gibraltar, Palma, Algiers, Tunis, Malta, Catania, Athens, Alexandria, Cairo (Helwan), Nicosia, Beirut, Baghdad, Lenkoran, Busrah, Meshed, Quetta, Lahore, Simla, Leh, Dehra Dun, Hang Kow, Tiensin, Shanghai (Zi-Ka-Wei), Chemulpo, Nagasaki, Kioto, Tokio, Miyako.

In Zone 6, latitude 20° to 30° N., there are 32 stations, viz.: Honolulu, Zacatecas, Leon, Galveston, New Orleans, Havana, Key West, Nassau, Bahamas; Puerto de Orotava, Las Palmas, Canary Is.; Insalah, Dakhla Oasis, Wadi Halfa, Suez, Aswan, Bushire, Jask, Kurrachee, Hyderabad, Sind; Jaipur, Nagpur, Allahabad, Calcutta, Gauhati, Cherrapunji, Shillong, Akyab, Phu Lien, Moncay, Hong Kong, Taihoku, Formosa; Naha.

In Zone 7, latitude 10° to 20° N., there are 39 stations, viz.: Zapotlanejo, Morelia, Salina Cruz, Mexico, Oaxaca, Belize, San Salvador, Jamaica (Negril Point), Port au Prince, St Croix (Christiansted), Grenada (Richmond Hill), Barbados, St Vincent, C. Verde Is.; San Tiago, C. Verde Is.; Bathurst, M'Carthy Island, Ségou, Timbouctoo, Sokoto, Kontagora, Maiduguri, Katagum, El Obeid, Khartoum, Aden, Bombay,

[1] Monthly Meteorological Data for ten-degree squares in the Atlantic and Indian Oceans. *K. Ned. Met. Inst.* No. 107a. Utrecht, 1914.

Mysore, Kodaikanal, Madras, Waltair, Port Blair, Rangoon, Saigon, Nhatrang, Bolinao, Vigan, Manila, Ormoc, Guam, Ladrone Is.

In Zone 8, latitude 0° to 10° N., there are 29 stations, viz.: Fanning Island, Culebra, El Peru, Guayanavieja; Georgetown, Brit. Guiana; Paramaribo, Cayenne, Conakry, Sierra Leone, Cape Coast Castle, Accra, Lagos, Zungeru, Libreville, Yola, Kafia Kingi, Wau, Entebbe, Nandi, Cothin, Colombo Observatory, Candy, Medan, Penang, Singapore, Sandakan, Tagbilaran, Surigao, Yap, Uyelang.

In Zone 9, latitude 0° to 10° S., there are 23 stations, viz.: Malden Island, Manaos, Fernando Noronha, Sainte-Croix-des-Eshiras, Loango, Tabora, Nairobi, Zanzibar, Lamu, Seychelles, Padang, Batavia, Pontianak, Passeroean, Kajoemas, Koepang, Ambon, Daru, Port Moresby, Rakuranga, Rendova, Tulagi, Brit. Solomon Is.; Ocean Island.

In Zone 10, latitude 10° to 20° S., there are 23 stations, viz.: Alofi, Niue Is.; Tahiti (Makatea), Puerto de Arica, Arequipa, Sucre, Bolivia; Araguaya, Cuyaba, St Helena, Salisbury, Rhodesia; Zomba, Tamatave, Tananarivo, Cocos Keeling Is., Christmas Is., Derby, Hall's Creek, Port Darwin, Daly Waters, Mein, Georgetown, Queensland; Samarai, La Kolle, Suva, Fiji.

In Zone 11, latitude 20° to 30° S., there are 27 stations, viz.: Rarotonga (Avarua), Mataveri, Easter Island; Punta Tortuga, Coquimbo; Puerto de Antofagasta, Goya, Porto Alegre, Curityba, Rio de Janeiro, Windhuk, Johannesburg, Kimberley, Pretoria, Durban, Lorenzo Marques, Mauritius, Onslow, Peak Hill, Nullagine, Laverton, Alice Springs, William Creek, Boulia, Mitchell, Rockhampton, Brisbane, Noumea, Norfolk Island.

In Zone 12, latitude 30° to 40° S., there are 18 stations, viz.: Juan Fernandez, Punta Angeles, Valparaiso; Santiago, Cordoba, Buenos Ayres, Montevideo, Cape Town, Perth (Observatory), Katanning, Coolgardie, Eucla, Streaky Bay, Adelaide, Melbourne, Bourke, Sydney, Lord Howe Island, Auckland.

In Zone 13, latitude 40° to 50° S., there are 9 stations, viz.: Chatham Island, Punta Corona (Puerto de Ancud), Punta Galera, Bahia Blanca, Hobart, Launceston, Dunedin, Christchurch, Wellington.

In Zone 14, latitude 50° to 60° S., there are 5 stations, viz.: Islota de los Evangelistas, Puerto de Punta Arenas, Punta Dungeness, Cape Pembroke, South Georgia (Grytviken).

In Zone 15, latitude 60° to 70° S., there is 1 station, viz.: South Orkneys.

In Zone 16, latitude 70° to 80° S., there are 3 stations, viz.: Framheim, Cape Evans (McMurdo Sound), Cape Adare.

The information asked for by the Solar Commission included Mean Daily Pressure, Means and Extremes of Temperature, and Rainfall. In dealing with pressure at stations of the second order it is usual not to introduce the reduction to sea-level but to give the mean of the pressure-values actually observed and leave any student of the figures to make the reduction for himself if he thinks it to be necessary. For the purpose, however, of comparing the pressure data from stations in the same region, reduction to a common level is necessary, and the carrying out of this operation for all the stations is a very useful process of examination of the data. In the *Réseau Mondial*, therefore, pressure at mean sea-level is given in a separate column in addition to pressure at station-level.

The process of deducing the mean pressure and the mean temperature for the day from the observations at fixed hours is a very intricate one when stations in all parts of the earth are used together. The formulae that have been employed by the several meteorological offices in preparing their data for publication are set out for each several station in a list of the stations following the results for each month and for the whole year, so that the reader can, by reference to this list, satisfy himself as to the meaning of the figures which are inserted in the columns of the tables.

Where normals are available, differences from normal have been inserted in separate columns for pressure, temperature and rainfall. Normals for many of the stations in the British Empire have been specially compiled for the purpose.

UNITS

The units in which the values are expressed in the original publications or forms are either British units, viz., inches of mercury, degrees of Fahrenheit's scale and inches of rainfall, or Continental units, viz., millimetres of mercury, degrees of the centigrade scale and millimetres of rainfall. For the *Réseau Mondial*, pressures have been given in millibars, temperatures in the absolute scale based on the centigrade degree[1], and rainfall in millimetres.

A word or two in explanation of this conclusion may not be out of place. When the figures for the whole world are before you and one half of them are in one system of units and the other half in another system, some consideration of the claims of the one or the other to finality becomes inevitable. So long as one is concerned merely with the comparison of the meteorological data one with another any system will do, provided the same is used for all; but as soon as pressure is regarded as having dynamical significance and temperature a thermodynamical significance the expression of the meteorological data in suitable units takes on an altogether different aspect.

The advantage of using dynamical units for expressing pressure becomes evident when we consider the ordinary practice of a large number of observers. It has become involved in a considerable number of processes of which the meaning has become obscured. An observer at sea, for example, reads in "inches," and sets down the temperature of the mercury. If he corrects the reading for temperature it remains in "inches," corrected for latitude it is still in "inches." Though "inches" have been used at every stage the reading has never really been inches at any stage, and after all it is not the inches of the atmosphere that we want to express but the pressure of the atmosphere. An accomplished philosopher when he knows the inches can compute the pressure because he knows the value of gravity in latitude 45° and the density of mercury at the freezing point, but that excludes the observer from all ideas of pressure as a dynamical quantity in the meteorology of the globe. If therefore we wish to associate the observer with us in regarding the atmospheric changes as the results of dynamical processes the first step is to express the pressure of the atmosphere as pressure and not as inches. Other steps may be necessary before the desirable object is achieved, but in any case this first step is indispensable.

[1] In order to make a distinction between the strictly absolute thermodynamic scale of temperature and the scale of centigrade degrees measured from a zero 273° below the normal freezing point of water which is referred to, the name *tercentesimal scale* has been proposed for the latter.

If we are agreed that a pressure-unit is required, the choice of the particular unit the millibar, 1000 dynes per square centimetre, is not likely to give rise to much discussion; the reasons have been set out already elsewhere.

As for the use of the absolute temperature on the centigrade scale, which differs only from the centigrade temperature by the addition of 273°, the first point to be noticed is that since the laws of thermodynamics have been formulated the numerical value of the temperature on the absolute scale has acquired very great significance. First, from its practical agreement with the scale of the air-thermometer the absolute temperature is one of the indices of the density of air, which is a vital consideration in many practical affairs. For the arm-chair meteorologist the air goes up the chimney by the absolute scale, for the aeronaut the balloon and the airship float according to the absolute scale, the aeroplane's rate of ascent and the limit of its height are governed by the absolute scale. And taking into consideration other phenomena than the density of air let us note for the observer that terrestrial radiation goes by the fourth power of the absolute scale; cloud formation depending upon dynamical cooling is governed by the absolute scale, for the absolute scale governs all the transformation of heat into other forms of energy.

Any one, therefore, who wishes to deal with any other aspect of meteorology than the simple comparison of data must be familiar with what is meant by the numerical expression of temperatures on the absolute scale. Any meteorologist who wishes to "intrigue" his friends in the wide outlook of the science of meteorology could find no better text for his discourse than the expression of the freezing point of water as 273° instead of 0° or 32°.

An enormous incidental advantage is that there are no negative temperatures and so no minus signs in a table of temperatures. Throughout the whole of the pages of the *Réseau Mondial* plus and minus signs occur only in the columns of "differences from normal," except in two instances in the column for height which indicate that the stations are below sea-level. That fact has more than a conventional significance. According to modern views of the meaning of the temperature of a gas as depending on the motion of its molecules, the change of sign at the freezing point of water is a stumbling-block to all but skilled physicists. The use of absolute temperatures is not a question of saving trouble to the more accomplished meteorologists but an opening upon an avenue by which the ideas of modern physics may become automatically a commonplace of all students of meteorology.

Whether the temperature on the absolute scale should be according to the Fahrenheit or the centigrade degree is a much more debatable question. In the one case the freezing point is 491°, and 500° on that scale is 41° F.; so a rise of 5° means 1 per cent. change in the density of air just as a change of 10 millibars in pressure means a change of 1 per cent. in the density: for the processes of mental arithmetic which any one must employ when he is thinking about the atmosphere that is much more convenient than the absolute temperature on the centigrade scale on which the freezing point of water is 273 a. To get a round number we have to go up to 300 a, which is rather high, or down to 250 a, which is very low. But so many tables of physical constants are computed and published according to the centigrade scale, which is the scale used in the teaching of physics in all our schools and universities, that it is too late now to work with the Fahrenheit scale. Moreover, the temperature of the freezing point is not so accurately represented by 491 Fahrenheit degrees above the absolute zero as it is by 273 centigrade degrees.

ADDITIONAL INFORMATION

When the data for the 392 stations came to be printed, arranged on the general plan of a page for each zone of 10° of latitude, it was found that the material for a month occupied thirteen pages. In a book of reference of this kind it is convenient to have corresponding data on corresponding pages, and the fourteenth page of each month was accordingly filled by additional data. Among the facts and figures which it is desirable to bring into relation with the climatic data for the globe we may single out first the state of the north polar ice, concerning which information is regularly collected by the Danish Meteorological Service, and secondly the winds within the tropics. These last are a subject about which sweeping general statements are frequently made in the discussion of the general circulation of the atmosphere, but very little precise information has been compiled. The stations selected were St Helena in latitude 16° S., which gives the velocity and direction of the south-east trade-wind at a height of 632 metres taken from the records of a Robinson anemometer maintained by the Meteorological Office; Suva, Fiji, in nearly the same latitude, 18° S., but opposite in meridian, for which also the velocity and direction of the wind are available; Malden Island, Tulagi, and Ocean Island between these two stations and the Equator, with Fanning Island and Sierra Leone just on the other side of the Equator. The contrast between St Helena and Suva is very striking. At the former the wind is persistently steady from the south-east, at six or seven metres per second with no calms; for the latter we have the following summary for the year 1911, based on the observations at 9 a.m.:

Table of Wind-directions at 9 a.m. at Suva, Fiji, in 1911

Month	N.	N.E.	E.	S.E.	S.	S.W.	W.	N.W.	Calm
January .	6	3	1	5	...	4	...	...	12
February .	4	5	1	5	2	2	...	...	9
March .	4	6	1	2	...	2	...	3	13
April .	7	12	1	1	...	...	1	4	4
May .	2	15	6	2	...	2	...	3	1
June .	3	6	3	1	1	1	...	5	10
July .	1	5	3	4	1	1	1	4	11
August .	4	6	4	5	...	1	...	4	7
September	2	10	8	3	...	3	2	2	...
October .	1	10	7	4	3	2	...	1	3
November	...	16	6	3	1	...	...	...	4
December	9	10	9	2	...	1	...	...	...
Year	43	104	50	37	8	19	4	26	74

It is difficult to reconcile these figures with the conventional explanation of the trade-wind exhibited at St Helena. If the explanation finds its most effective illustration in latitude 16° S. 6° W., why is there no corresponding effect in 18° S. 178° E.? We should expect a south-east trade-wind at the one, just as confidently as at the other, but the prevailing direction of the wind at Suva is north-east and the next entry as regards numerical frequency is that of calms. Over both places the sun is vertical twice in the course of the year, and other circumstances are very similar, yet the *régime* of the winds is quite different.

The theory of the origin of the trade-winds received a good deal of attention recently when Teisserenc de Bort found the predicted counter-trade flowing in the opposite direction above the trade-wind. This was in the northern hemisphere, and a south-west wind was found over the north-easter. But since the exploration of the air with pilot-balloons has become general we have learned to expect a south-west wind over a north-easter in our own latitudes, and in that case we cannot attribute it to the ascent of air at the terminus of the north-east current because such currents generally have no terminus on our maps.

There is evidently here a great opportunity for some one to examine once more the facts which the theory of the trade-winds is designed to explain.

The Results

With the completion of these tables, which summarise about 1,500,000 observations, the inevitable has happened. There are about ten columns of meteorological figures on each page, and the whole comprises 392 × 10 × 13, about 50,000 values altogether. I know of no meteorological situation that is better calculated to daunt the ardour of an open-minded searcher after the truth about the weather than to present to him 50,000 numbers and give him to understand that the solution of his problems, or at any rate the beginning of it, lies therein. And yet it is a step in the right direction and we have to find the means of turning it to good account. I cannot pretend to have done that; the figures have only come from the printer within the last few days, and though it is much easier to make a survey of them now that they are in type than it was when they were in manuscript, still—

> The world is all before us, where to choose,

and the choice is bewildering. One naturally runs to maps, and accordingly thirteen maps have been prepared, one for each month and one for the year. The maps show the departures of pressure and temperature from the normals, the former by lines, the latter by figures. Many others of the values might be represented by maps, but it is necessary to set a limit. The projection of the globe which is used for these maps is the one out of which we are accustomed at the Meteorological Office to construct what we call octagonal globes. The surface is divided into five pieces: A cylindrical strip representing the equatorial region between the tropics, two developed conical pieces for the temperate zones, and two circular discs for the regions north and south of the Arctic and Antarctic circles respectively. When mounted on card and made up they form a solid of octagonal section through the polar axis and circular section at right angles thereto. The projection is adjusted so that the scale is everywhere within 5 per cent. of uniformity[1].

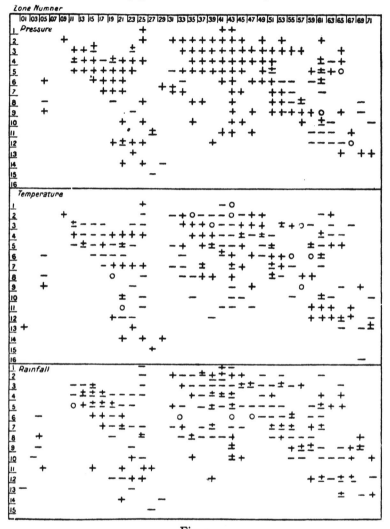

Fig. 3

Chart showing the 10° squares in which the deviations from the normals of pressure, temperature, and rainfall in July 1911 were above or below the average. The numbers on the left-hand side indicate the zones of latitude of pp. 22–23, those at the top the steps of ten degrees in longitude from 180° W., the first of which is numbered 01 and the next 03 and so on. The sign + indicates that the value at both stations is above normal, – that both are below, ± one above and one below, o no appreciable difference from normal.

These maps show easily enough the limits of the region within which the abnormally high temperatures of the summer of 1911 were experienced, and

[1] This method of mapping modified in its detail at the suggestion of Prof. V. Bjerknes was adopted by the International Meteorological Committee in 1921. Report of the Eleventh Ordinary Meeting, M.O. Publication, 248.

they show also the relation of the high temperature to the high pressure, a relation sufficiently close to dispose of the impression which once governed meteorological theory that a circumscribed area of high temperature is the cause of and therefore the accompaniment of low pressure.

Zone	Number of Stations	Jan.	Feb.	Mar.	April	May	June	July	Aug.	Sept.	Oct.	Nov.	Dec.
1	3	-3.9	-6.4	-3.0	-5.0	-1.6	-0.6	+2.0	+1.3	+0.0	-3.5	-4.0	+3.0
2	20	+1.1	-5.0	+2.3	-3.0	-1.4	+0.7	+3.8	+1.9	-0.3	+0.5	-4.0	-0.2
3	31	+3.5	-7.6	-0.3	-0.6	+0.0	+1.8	+3.3	+0.5	-0.3	-0.3	-3.4	+1.7
4	35	+0.5	+3.0	+0.2	+0.9	-0.3	+0.9	+1.7	+0.5	+0.4	+1.2	-0.7	+1.3
5	40	+0.5	+2.7	+0.0	+0.4	+0.6	+0.7	+1.0	-0.2	+0.4	+0.5	+0.4	+0.9
6	22	-0.7	+1.4	-0.0	-0.1	-0.1	+0.4	+0.7	-0.8	-0.0	+0.6	+0.3	-0.1
7	17	-0.3	+1.3	+0.6	+0.5	+0.2	+1.1	+1.1	+0.4	+0.4	+1.3	+0.7	+0.1
8	13	-0.4	+0.1	+0.0	+0.0	-0.3	+0.5	+0.3	-0.2	-0.2	+0.7	+0.2	-0.3
9	12	-0.4	+0.2	+0.3	-0.0	-0.3	+0.2	+0.4	+0.1	+0.1	+1.2	+0.3	-0.2
10	13	-0.6	+0.2	+0.5	+0.7	+0.1	+0.9	+1.2	+0.5	+0.7	+1.4	+0.5	+0.6
11	11	-0.7	-0.4	+0.0	-0.9	-0.4	+1.8	-0.2	-0.2	-0.4	+0.8	-0.8	-1.0
12	16	-0.6	+0.0	+1.2	1.1	+0.0	+2.9	-0.2	+0.8	+0.2	+1.9	-0.6	-1.9
13	5	+2.1	+3.6	+3.0	-3.8	+5.3	-1.6	+4.0	+3.4	+2.4	+1.6	-4.9	-8.0
14	3	-2.6	+11.0	+6.1	+0.4	+7.4	+4.2	+1.9	-6.6	+5.9	+2.4	-0.5	+5.5
15	1	+2.8	+3.7	-1.7	+9.2	+18.2	+1.2	-0.6	-4.9	+3.6	+5.2	+3.8	+10.2
16	1	-1.1	-1.0	-3.6	2.1	-0.3	-3.5	-2.8	+2.6	-3.5	-5.8	+8.9	+10.8

Fig. 4

The means of the pressure departures of the stations in successive zones of latitude for each month of the year 1911.

There are some interesting examples of the concentration of high or low pressure at two antipodal points, but the maps do not, at first sight, disclose the secret of the peculiarities of the year's weather.

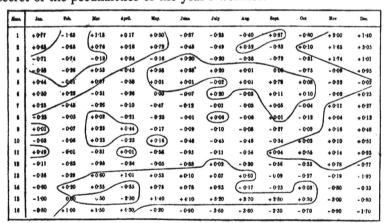

Zone	Jan.	Feb.	Mar.	April	May	June	July	Aug.	Sept.	Oct.	Nov.	Dec.
1	+0.77	-1.63	+1.13	+0.17	+0.50	-0.87	-0.23	-0.40	+0.87	-0.60	+2.00	+1.40
2	+0.65	-0.65	+0.76	+0.16	+0.73	-0.48	-0.49	+0.52	-0.53	+0.10	+1.65	+3.05
3	-0.71	-0.74	-0.19	+0.84	-0.16	+0.20	-0.20	-0.36	-0.72	-0.21	+1.74	+1.01
4	-0.58	-0.26	+0.53	+0.45	+0.56	+0.26	+0.20	+0.01	0.00	-0.75	-0.09	+0.95
5	+0.44	-0.51	+0.27	-0.38	+0.01	+0.01	-0.02	+0.01	+0.78	+0.08	-0.32	-0.02
6	+0.50	+0.22	-0.31	-0.26	0.00	-0.07	+0.20	-0.03	+0.11	+0.10	-0.03	+0.23
7	+0.25	-0.45	-0.26	-0.10	-0.47	-0.12	-0.01	-0.03	+0.05	-0.04	+0.11	+0.27
8	-0.23	-0.05	+0.02	-0.21	-0.25	-0.01	+0.06	-0.06	-0.01	-0.13	+0.04	+0.12
9	+0.01	-0.07	+0.33	+0.44	-0.17	-0.09	-0.10	-0.08	-0.27	-0.09	+0.16	+0.48
10	-0.03	-0.06	+0.23	-0.23	+0.16	-0.46	-0.45	-0.48	-0.34	-0.03	+0.10	+0.51
11	+0.49	-0.01	-0.31	+0.01	-0.56	-0.95	-0.11	-0.54	+0.04	+0.55	+0.14	+0.93
12	-0.11	-0.83	-0.94	-0.24	-0.05	-0.88	+0.02	-0.30	-0.56	-0.33	+0.78	-0.77
13	-0.36	-0.29	+0.60	+1.01	+0.53	+0.10	+0.07	+0.60	-0.09	-0.27	-0.19	-1.90
14	-0.60	+0.20	+0.35	-0.35	+0.75	+0.75	+0.93	-0.17	-0.23	+0.03	-0.80	-0.33
15	-1.00	0.50	0.50	-2.30	+1.40	+4.10	+3.20	+3.70	+2.80	+0.30	-3.00	-0.80
16	-0.80	+1.00	+1.50	+4.30	-0.20	-0.90	-3.60	-3.60	-2.30	-0.70	-0.90	-1.70

Fig. 5

The means of the temperature departures of the stations in successive zones of latitude for each month of the year 1911.

By way of forming some general ideas of the state of things in different parts of the globe I have marked with + or − on the accompanying diagram (Fig. 3) the squares which showed pressures, temperatures and rainfalls above or below normal in the month of July. The agglomeration gives a rough idea of the distribution of land and water and the extent of the regions of high and

TEMPERATURE. *Absolute Maximum in each Zone*

Zone	January	February	March	April	May	June	July	August	September	October	November	December
1	Mehavn 277·9	Gjesvaer 278·1	Upernivik 278·7	Gjesvaer 277·8	Gjesvaer 290·1	Vardö 288·4	Gjesvaer 297·8	Mehavn 296·0	Gjesvaer & Vardö 298·0	Mehavn 281·7	Upernivik 283·9	Gjesvaer 279·8
2	Thorshavn 284·0	Thorshavn 283·2	Bergen 283·6	Bergen 290·2	Petrozavodsk 300·2	Troitsko-Petcherskoe 303·3	Olekminsk 304·4	Iakoutsk 304·1	Fort Simpson 298·6	Fort Simpson 291·3	Thorshavn 286·0	Thorshavn 284·8
3	Valencia & Greenwich 284·1	Aberdeen 286·9	Potsdam 296·3	Minnedosa 299·9	Kamloops 304·4	Saratov 311·0	Kazan 312·4	Greenwich 310·8	Potsdam 307·7	Kamloops 304·0	Greenwich 288·0	Kamloops 288·0
4	N. Platte 293·6	Madrid 294·9	N. Platte 299·7	Tachkent 305·0	N. Platte 308·0	N. Platte 313·0	Toronto 312·6	Bucharest 312·2	Madrid 312·3	Bucharest 302·2	Chicago 296·3	Marseilles 291·0
5	Nashville & S.L.Obispo 298·6	Abilene 304·7	Abilene 307·4	Lahore 311·8	Lahore 319·9	Lahore 320·2	Baghdad 320·1	Baghdad 318·1	Baghdad 318·4	Lahore and Tunis 313·5	Lahore 303·8	Alexandria 299·0
6	Wadi Halfa 307·0	Wadi Halfa 310·0	Nagpur 312·2	Wadi Halfa 318·0	Hyderabad 319·7	Insalah 320·5	Insalah 323·0	Insalah 321·0	Insalah 318·0	Insalah 317·0	Hyderabad 308·8	Wadi Halfa 305·0
7	Khartoum 312·1	Khartoum 314·0	Katagum 316·9	Timbuctoo 319·0	Timbuctoo 319·0	Timbuctoo 319·0	Khartoum 316·3	Timbuctoo 315·0	Timbuctoo 318·5	Timbuctoo 317·0	Timbuctoo 313·0	M‘Carthy Island 311·3
8	Wau 311·0	Zungeru 312·4	Yola 313·6	Yola 313·0	Yola and Zungeru 310·2	Yola, Wau & Penang 308·0	Medan 308·8	Medan 308·1	Wau 309·0	Wau 310·0	Yola 309·1	Zungeru 310·2
9	Sainte-Croix 307·0	Sainte-Croix 307·8	Sainte-Croix 310·0	Sainte-Croix 309·3	Sainte-Croix 307·0	Lamu 309·7	Malden Is. 306·3	Malden Is. 306·9	Lamu 310·8	Malden Is. 306·4	Tulagi 308·0	Malden Is. 308·0
10	Hall's Creek 315·8	Hall's Creek 313·4	Hall's Creek 314·9	Hall's Creek 312·4	Derby 307·9	Puerto de Arica 307·2	Daly Waters 306·3	Derby 307·9	Hall's Creek 311·2	Derby 314·8	Hall's Creek 315·2	Derby 316·2
11	Onslow 318·9	Onslow 316·2	Nullagine 315·7	Onslow 312·3	Nullagine 307·8	Boulia 304·1	Boulia 305·2	William Creek 306·3	Lorenzo Marques 313·7	Boulia 314·8	William Creek 318·0	William Creek 317·7
12	Coolgardie 315·6	Eucla 315·1	Eucla 312·6	Coolgardie 309·7	Bourke 305·7	Cape Town 300·2	Eucla 298·8	Eucla 305·8	Bourke 307·2	Eucla 313·8	Bourke 314·2	Bourke 316·3
13	Hobart 305·1	Launceston 303·6	Christchurch 302·9	Christchurch 299·1	Hobart 292·4	Christchurch 292·2	Punta Galera 291·8	Hobart 292·4	Christchurch 300·3	Hobart 301·2	Launceston 301·4	Launceston 299·3
14	Cape Pembroke 292·4	Punta Dungeness & Punta Arenas 293·6	Punta Dungeness 293·8	Los Evangelistas 288·8	Los Evangelistas 290·0	Los Evangelistas 291·2	Los Evangelistas 289·8	Los Evangelistas 289·8	Los Evangelistas 288·0	Punta Arenas 291·5	Punta Dungeness 288·8	Punta Dungeness 296·4
15	South Orkneys 277·5	South Orkneys 281·0	South Orkneys 281·5	South Orkneys 276·4	South Orkneys 277·3	South Orkneys 274·6	South Orkneys 275·8	South Orkneys 273·4	South Orkneys 275·6	South Orkneys 281·2	South Orkneys 275·2	South Orkneys 277·1
16	Cape Evans 273·5	Cape Evans 273·6	Cape Adare 270·8	Cape Adare 267·4	Cape Adare 268·6	Cape Evans 266·3	Cape Evans 262·1	Cape Evans 263·7	Cape Adare 266·3	Cape Adare 269·7	Cape Adare 272·4	Cape Adare 276·9

TEMPERATURE. *Absolute Minimum in each Zone*

Zone	January	February	March	April	May	June	July	August	September	October	November	December
1	Upernivik 242·9 / Verk-hoiansk **209·2**	Upernivik 236·0 / Verk-hoiansk **214·8**	Upernivik 238·0 / Verk-hoiansk **223·8**	Upernivik 245·5	Upernivik 257·0 / Verk-hoiansk 247·3	Upernivik 269·0	Upernivik 270·0	Upernivik 270·3 / Verk-hoiansk 269·6	Upernivik 264·5 / Verk-hoiansk 256·9	Upernivik 262·5	Upernivik 249·0 / Verk-hoiansk **223·5**	Spitsbergen 233·8
2	Blagovech-tchensk Priisk 221·9			Markovo 234·6	Omsk 260·1	Gothaab 267·6	Grimsey 272·8	Blagovech-tchensk Priisk 271·5	Tygan Ourkan 261·9	Doudinka 239·3 / Blagovech-tchensk Priisk 247·0	Blagovech-tchensk & Kirensk 230·2	Iakoutsk **213·2**
3		Kirensk 221·7	Kirensk 230·9	Chipewyan 244·1	Bismarck 265·2	Tygan Ourkan 267·4	Tygan Ourkan 272·5	Bismarck 273·0	Port Arthur 268·6	Port Arthur & Cheyenne 262·4		Kirensk 224·7
4	Macleod 230·2	Ochiai 235·7	Ochiai 241·2	Ochiai 250·4		St Johns 273·0	Ochiai 277·0				Macleod 242·2	Macleod 235·8
5	Denver 245·8 / Galveston 265·8	Leh 248·3 / Shillong 271·7	Leh and Santa Fé 260·8	Modena 263·6	Santa Fé 271·9	Modena 275·2	Tunis 280·6	Leh 279·1	Modena 274·1	Denver 264·7	Santa Fé 250·8	Modena 245·8
6			Dakhla 276·0	Zacatecas 274·2	Leon 283·5	Zacatecas 280·6	Leon 285·3	Zacatecas 280·8	Leon 284·5	Zacatecas 278·4	Zacatecas 267·0	Zacatecas 272·4
7	Mexico 274·9	Mexico & Zapotlanejo 277·0	Mexico 278·6	Zapotlanejo 280·0	Mexico 281·4	Mexico 283·4	Mexico 281·2	Kodaikanal 281·6	Mexico 283·2	Mexico 282·0	Mexico 277·7	Mexico 275·2
8	Nandi 281·9	Nandi 281·9	Nandi 283·0	Nandi 283·0	Nandi 281·9	Nandi 282·4	Nandi 281·9	Nandi 280·8	Nandi 279·7	Nandi 281·9	Nandi 281·3	Nandi 280·2
9	Nairobi 280·2	Nairobi 278·6	Nairobi 277·4	Nairobi 283·3	Nairobi 283·6	Nairobi 280·2 / Hall's Creek 280·2	Nairobi 276·9	Nairobi 279·7	Nairobi 276·3	Nairobi 280·5	Nairobi 281·1	Nairobi 280·2
10	Arequipa 280·8	Arequipa 280·2	Arequipa 276·3	Arequipa 279·1	Salisbury 277·2		Salisbury 271·9	Salisbury 276·1	Salisbury 275·2	Arequipa 278·6	Arequipa 280·2	Arequipa 280·8
11	Punta Tortuga 281·0	Johannesburg 279·6	Johannesburg 279·3	Mitchell 273·0	Pretoria 271·9	Mitchell **265·8**	Mitchell 268·3	Mitchell 269·7	Mitchell 272·4	Pretoria 276·8	Mitchell 279·1	Mitchell & Johannesburg 282·4
12	Santiago 280·4	Katanning 279·4	Katanning 278·2	Santiago 274·1	Santiago & Montevideo 274·0	Montevideo 269·0	Santiago 270·5	Santiago 269·8	Santiago 273·4	Montevideo 272·0	Melbourne 275·7	Santiago 277·6
13	Christchurch 275·4	Christchurch 276·7	Christchurch 275·4	Launceston 274·9	Christchurch 270·0	Christchurch 270·1	Christchurch 268·9	Christchurch 271·2	Christchurch 271·6	Christchurch 272·1	Dunedin & Chatham Is. 274·1	Christchurch 275·9
14	Cape Pembroke 275·2	Punta Arenas 275·4	Punta Dungeness 273·6	Punta Arenas 273·4	Punta Arenas 271·0	South Georgia 264·4	South Georgia 262·7	South Georgia 261·7	South Georgia 260·4	South Georgia 263·7	South Georgia 268·4	South Georgia 268·9
15	South Orkneys 268·8	South Orkneys 268·0	South Orkneys 262·3	South Orkneys 256·5	South Orkneys 251·2	South Orkneys 253·0	South Orkneys 248·7	South Orkneys 255·0	South Orkneys 245·0	South Orkneys 249·0	South Orkneys 261·4	South Orkneys 263·8
16	Cape Evans 257·9	Cape Evans 251·6	Cape Evans 250·7	Framheim **225·0**	Framheim **222·4**	Framheim **214·8**	Framheim **219·0**	Framheim **214·5**	Framheim **220·0**	Framheim **232·8**	Framheim 245·0	Framheim 256·2

low values; both signs are given when the two stations in the same square give opposite results. The peculiarity shown in this diagram is the large predominance of positive signs in the chart for pressure. It must be supposed that as the integral of the pressure over the whole globe is simply the total weight of the atmosphere, high pressure in one locality must be exactly compensated with low pressure in others. That is evidently not the case with the values set out for July.

Various suggestions may be made to account for this remarkable result. First, the values are all for land stations and there are vast areas of sea over which so far as we know at present the pressures may be below normal; that suggestion is, however, not generally borne out in the case of the squares for which values have been given by the Netherlands Office and are included in Fig. 3. Next, it may be remarked that the pressure-values are reduced to sea-level with an assumed value for the temperature of the column between the station-level and sea-level. Errors may lurk in the selection of this temperature which eventuate in too high values.

These and possibly other suggestions will require examination before we can accept the alternative conclusion that the total weight of the atmosphere varies from year to year.

As a contribution to the subject I have obtained the mean values of pressure for all the stations in the same zone of ten degrees of latitude and set them out in Fig. 4.

The preponderance of high pressures is most striking in June and July, though it is by no means absent in other months. November, December and January are indeed the only months which seem to provide an ordinary balance.

Corresponding values have been taken out for temperature and are exhibited in Fig. 5. There is in this no corresponding suggestion of the predominance of warmth or cold. What one notices in this case is the association of areas of low temperature in successive months and the patch of high temperature in the north temperate zones lasting from April to October. This includes our warm period, but oddly enough Zone 3 (50° to 60° N.), which includes our islands, was not in the aggregate above normal in temperature during the summer months; Zone 4 next to it clearly was.

There is a suggestion about this diagram that a cold area once formed is not easily got rid of. It may move to other zones, but it is not annihilated. The same may be said with rather less insistence of the warm areas. There are a number of local ones, but most of the warm areas also are connected.

Extremes of Temperature in each Month in each Zone

Finally, it may be of interest to quote the highest and lowest temperatures that are found in the records for each zone in the course of each month. They are set out in the tables on pp. 30, 31. Figures in italics show the extremes for the zone during the year, figures in black type show the extremes for the known meteorological world for each month.

From these, with due regard, of course, to their successors, the reader may, if he please, select the climate of his taste.

CLOUD FORMS

PLATE XXV

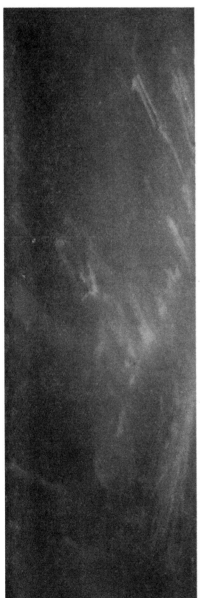

Fig. 6. Wisps of Cirrus in a blue sky.

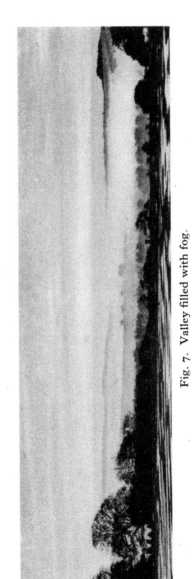

Fig. 7. Valley filled with fog.

The illustrations of cloud forms which are reproduced here are from the collection of C. J. P. Cave of Stoner Hill, Petersfield, with the exception of Fig. 13, which is from the Cloud-Atlas of Signor Luigi Taffara.

PLATE XXVI

Fig. 8. Alto-cumulus.

Fig. 9. Strato-cumulus seen from an aeroplane.

CLOUD FORMS

PLATE XXVII

Fig. 10. Cumulo-nimbus (near).

Fig. 11. Cumulo-nimbus (far-off).

Fig. 12. Lenticular Cloud of South Coast.

Fig. 13. Lenticular Cloud of the district of Mt Etna.
Contessa del Vento. (From Taffara's Cloud-Atlas)

4. THE WEATHER OF THE WORLD

A Lecture delivered at Uppingham School, 21 October, 1920

You will probably wonder why I chose for this lecture a title so comprehensive as " The Weather of the World." For the short hour of an afternoon's talk, the weather of Uppingham or of the County of Rutland might have been more than enough and more directly interesting, or, going beyond that limit, the weather of the British Isles has sufficient points of interest to satisfy the most ambitious lecturer.

It is not ambition but a modest diffidence that impels me to ask your attention to the wider view. If I felt that I could really tell you in one lecture all that you would like to know about the weather of Uppingham I would not hesitate to do so. There are people who have the reputation of knowing all about the weather of their own locality. There is a mythological meteorological personage known in history as " The Shepherd of Banbury," who is the personification of local weather-wisdom. He gave a code of rules for foretelling the weather. I have no doubt that Uppingham has its wise shepherd no less renowned, among those who know him, than he of Banbury. If he could be persuaded to follow the example of his brother of Banbury and write down his rules for foretelling the weather you would find that even the best of them would have notable exceptions, and when you look at the weather from the scientific point of view, as some of you may be disposed to do (that is to say, when you want to know all about it), you will wish more than anything else to find out what the exceptions mean.

Some thirty years ago I had to set an examination paper for a school examination in elementary hydrostatics and among the questions was one to "describe the barometer and explain its use." No meteorology was intended; but one of the answers which I got in return was "a barometer is an instrument shaped like a banjo. It is used to foretell the weather but it cannot foretell it for long." That particular schoolboy deserved his share of the reverence which is due to youth. What he says about the banjo is an overhasty generalisation; but it is good natural philosophy to remember what the barometer cannot do while reciting what it can do.

For the intervening thirty years I have been trying to find out why the barometer cannot foretell the weather for long, and what is the meaning of the exceptions to its rules. I have come to the conclusion that if you want to know all about the weather of Uppingham or any other place on this earth you had better begin by looking at the weather of the world as a whole. The weather of any locality is only an incident in a very complicated turmoil of air which forms what we call the general circulation of the atmosphere, and if we want to understand any part of it we must have a working knowledge of the general system.

I will not explain in detail what is included in the word "weather" the world over. You doubtless have mental pictures of your own; but by way of reminding you that evaporation of water and its counterpart condensation of water-vapour in the form of clouds and rain are the most expressive features of weather let me show you a few natural pictures of those features which, let me say, are for the most part common to the whole world, only more or less frequent in one place than in another.

We will begin with the condensation of water-vapour at very high levels, from 25,000 to 30,000 feet high, in the form of the thread-like cirrus cloud (Fig. 6, plate xxv). I show you that because such clouds are the highest visible and I want your special attention later on to the atmosphere *beyond* them. The dappled cirro-cumulus which takes many forms is nearly as high. We pass over alto-stratus and show a specimen of alto-cumulus (Fig. 8, plate xxvi) as a cloud of middle height and come to lower clouds at about 3000 to 6000 feet, with an example of strato-cumulus seen from an aeroplane (Fig. 9, plate xxvi); it is the most frequent and most persistent of our own forms of cloud. The same picture shows some heavy cumulus cloud projecting upwards from the great cloud-layer. We then pass on to rain clouds and see an example of cumulo-nimbus representing a thunder-storm as seen from a distance (Fig. 11, plate xxvii) and another example of cumulo-nimbus (Fig. 10, plate xxvii), a thunder cloud photographed while the storm was nearly overhead.

I add two specimens which represent essentially local clouds; one of a ground-fog filling the valley (Fig. 7, plate xxv) beyond which the cumulo-nimbus was seen in the first of the two photographs of that type, and the other a curious turban-shaped cloud (Fig. 13, plate xxviii). It belongs to the neighbourhood of Mount Etna which is seen in the photograph and is known by the local name Contessa del Vento, the countess of the wind. It is a very peculiar form of cloud; a noteworthy point about it is its smooth outline. We have clouds in this country too with smooth outlines reminding one of airships. They are probably also associated with the disturbance of the air currents caused by the features of the land. An example of such a cloud near our south coast is shown in Fig. 12, plate xxviii. Clouds of similar type have been photographed at Danmarks-Havn, Greenland.

I shall direct the train of thought particularly to rainfall because rain means so much for the changes of energy which take place in the atmosphere in the sequence of evaporation of water and condensation of water-vapour. The energy of a pound of water vaporised is about 600 thermal units, equivalent to 834,000 foot-pounds, that is equivalent to lifting a ton through 400 feet. I have worked out a little sum which I ask you to persuade your students of physics to verify or correct for you and for me. My conclusion is that for every inch of rain in the year there is energy enough to work at the rate of about 12 horse-power per acre throughout the year, 8000 horse-power for every square mile. So that any place with forty inches of rain a year has enough energy supplied to the air above it to work at the rate of a quarter of a million horse-power to the square mile, day and night the year through. If we could only, somehow or other, harness the energy which is represented by the annual rainfall we should not be much concerned about the effect of a miners' strike upon industry.

Now keeping this idea of a vast store of energy in mind let us look at the distribution of annual rainfall over the land areas of the globe. Before doing so let us take a view of the land areas themselves and form an idea of their

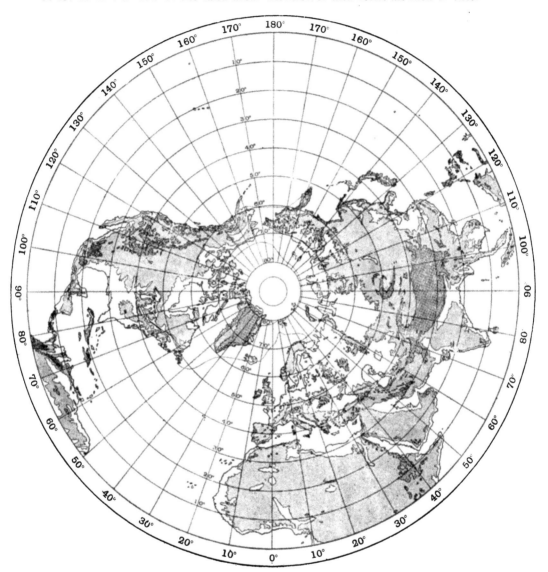

Fig. 14. Orographical Map of Northern Hemisphere.
The contours are at 200 and 2000 metres.
The ice zones over the sea for summer and winter are indicated by dotted lines.

distribution and features because they are of considerable importance in the distribution of rainfall. The map (Fig. 14) shows the sea-level line and the contours of 200 metres and 2000 metres; and I call attention particularly in the northern hemisphere to the high plateau of Greenland and to the roof of the world in Asia and the high land of North America. I leave you to make out for yourselves in the southern hemisphere (Fig. 15) the table-land of Africa, the Antarctic, and the curious low level of the interior of Australia.

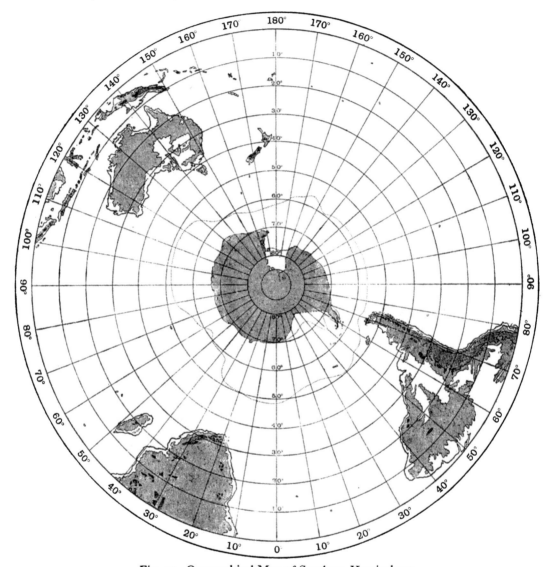

Fig. 15. Orographical Map of Southern Hemisphere.
The contours are at 200 and 2000 metres.
The ice zones over the sea for summer and winter are indicated by dotted lines.
The contours of the Antarctic Continent are unknown.

And for these figures 14 and 15 I want to have your attention particularly to the variation between the limits of ice in summer and winter in the northern and southern hemispheres; I ask you to think of the immense loss of heat which goes on automatically and is represented by the conversion of so much sea into ice, and the power of the sun to change it back again into water in the course of the year. The effect is even more conspicuous in the southern hemisphere where the area covered by ice and snow is doubled in the winter as compared with the summer. Every year, in both hemispheres combined, about one-eighth of the earth's surface is frozen over and melted again.

On the same outlines the average distribution of the year's rain (Figs. 16 and 17, plates II and III) is shown by lines of equal rainfall, or isohyets, drawn for rainfalls of 10, 20, 40, 80 and 120 inches. Ten inches of rainfall in the year are not as a rule sufficient to grow crops and so we see the deserts of the world mapped out by the areas enclosed within the 10-inch lines, the areas between the 40-inch lines and 20-inch lines are the regions where crops are secure, and the regions from 40 inches to 120 inches are those where special crops requiring a great amount of moisture can be grown without any special irrigation.

While these maps are before you I should like to draw your attention to the extensive co-operation which is necessary to obtain the result. In every part of the world reading a rain-gauge is a tedious daily operation and the map represents the repetition of that operation for many years at thousands of places all over the world. I know no other example so effective as a representation of the essential brotherhood of the whole world working towards a common aim.

The problem of the weather of the world is now before us and we can take a few steps in the direction of its general solution. The first is to recognise that the differences shown on the map are attributable to weather and the primary agent is temperature, warmth and cold, which in its turn is due to the heat of the sun's rays. The distribution is modified by the air currents which are set up by the local warming and cooling of the air, warmth in the equatorial regions and cold in the polar regions. It is also modified by the distribution of land and water.

Let us look, therefore, at the distribution of temperature in winter and in summer as represented by the isotherms for January and July in the northern hemisphere and for July and January in the southern hemisphere (Figs. 18–21, plates IV–VII). Note that the temperature gradually falls off from the equatorial regions to certain localities near the polar regions in either hemisphere. That is the only conclusion which you need draw at present though there are many other points of interest about these maps.

Bearing in mind that distribution as representing temperature at the surface let us now look into what happens in the upper air. Since 1896 a great investigation into the secrets of those regions has been going on by means of balloons carrying self-recording instruments to heights far beyond what can be reached either by manned balloons or by aeroplanes.

The first results of the investigation in this country are shown in Fig. 22, taken from a diagram in the Meteorological Office, and are very remarkable; the temperature falls comparatively uniformly until a certain limiting point is reached and then the fall of temperature ceases. This was so far contrary to expectation that for a long time the more cautious of our physicists refused to

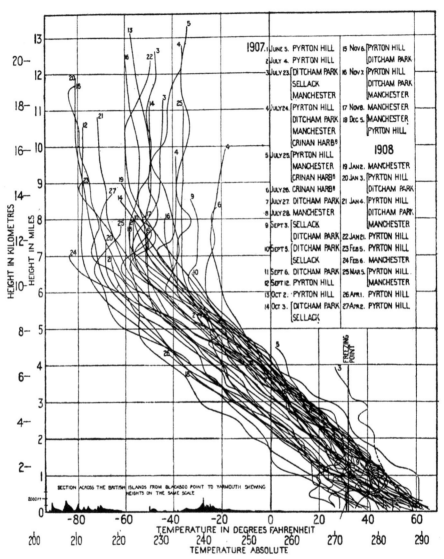

Fig. 22. Curves showing change of Temperature with height above sea-level obtained from balloon ascents 1907–8.

The separate curves represent the relation between temperature, in degrees Fahrenheit or on the absolute centigrade scale, and height in miles or kilometres in the atmosphere. The numbers marking the separate curves indicate the date of ascent at the various stations as shown in the tabular columns. The general aspect of the curves shows the great complexity of the temperature variations within the first two miles from the surface, and a very nearly uniform rate of fall of temperature above the two-mile limit until the isothermal layer is reached, at from six to eight miles. The difference of height at which the isothermal layer is reached, and the difference of its temperature for different days or for different localities, is also shown on the diagram by the courses of the lines.

believe it; yet it is true. The turning point marks the beginning of a new region called the stratosphere, below it is the troposphere. The temperature remains constant in each ascent from the base of the stratosphere to the greatest height hitherto attained, about 37 kilometres, 100,000 feet, but it is by no means the same in different ascents. There is as much difference of temperature between the extreme values observed at the tropopause, the boundary between the troposphere and the stratosphere, as there is at the ground.

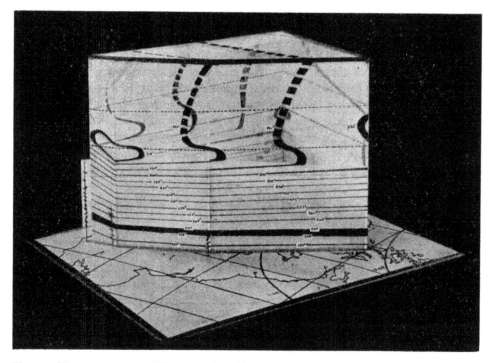

Fig. 23. Temperatures and Pressures in a block of Atmosphere 15 miles thick over a triangular portion of the British Isles.

From observations taken on 27 July, 1908. Block seen from the north-east. Isotherms are shown for each 5 a from 280 a to 215 a. The space between the isotherms of 270 a and 275 a is filled in; for other isotherms a thickness corresponding with $\frac{1}{2}°$ C. is covered. The beaded lines in the stratosphere are isobars for 200 mb. and 100 mb. respectively. The arrows on the standards face the wind as determined by observations with theodolites.

Many hundreds of such observations have now been obtained in various parts of Europe, in Canada, United States, in India, Java, East Africa, Spitsbergen, the Antarctic and over the Atlantic.

I cannot enter into the details of the analysis of all these ascents but I can show a photograph of a model which will enable me to illustrate the result of the observations (Fig. 23). You will see from the model which represents the distribution of temperature by isothermal surfaces, i.e. surfaces of equal temperature, that in the troposphere the temperature is stratified horizontally, but in the upper air layer after layer tilts its end up and the surfaces become more or less *vertical* instead of *horizontal*[1].

The important point to notice is that the lowest layer to turn up is of course the warmest of the vertical layers, and it is *at the north* and the turn up is later and later, and so colder and colder, further south. Hence, speaking generally, these approximately vertical columns of air are arranged with the coldest at the equator and the warmest at the pole. It depends how far you travel along horizontally, or nearly so, until you turn up. It becomes apparent that at very high levels the equator becomes the coldest place on earth and we have columns of cold air over the equator and warmer air as we approach the pole. Hence the distribution of temperature in the upper layer is the reflection, or the reverse, of that at the surface and extends through a great thickness.

Now we can pass from the distribution of temperature to the distribution of pressure. Looking at the atmosphere not from the bottom but from the outside, we may picture to ourselves a vast stratum of air many kilometres thick made up of columns each approximately uniform in temperature, the lower temperature being in the equatorial regions and higher as we go poleward. Each column will exert its pressure like a column of water and the colder air of the equator will produce a higher pressure than the warmer air of the polar regions. So we shall get a distribution of pressure high at the equator and low at the poles.

That is the distribution which was found forty years ago by computation from the pressure at the surface and the distribution so mapped out is justified by the recent knowledge of the upper air derived from the modern observations of temperature. The representation is shown by maps of isobars at the level of 4000 metres for the northern and southern hemispheres in January and July (Figs. 24–27, plates XIV–XVII).

It must now be noted that to a distribution of pressure such as that represented there corresponds a flow of air along the isobars in accordance with a law known as Buys Ballot's law, which asserts that if you stand with your back to the wind in the northern hemisphere the pressure is low on the left hand and high on the right, and we have reason to think that the law as applied to the upper air is a strict flow along the isobars with a current proportional to the closeness of the lines[2].

Hence the general circulation at the level of 4000 metres is circulation from

[1] The model which is shown in Fig. 23 represents the distribution of temperatures over Britain on one particular day—27 July, 1908. At Uppingham a cardboard model was exhibited representing the average distribution of temperatures over the globe by surfaces which started north and south from the equator and suddenly turned upward or came against the surface. Since the lecture the representation has been elaborated to what is shown in Fig. 47, and that figure should be referred to for what follows.

[2] The statement here made that in the upper air the flow is strictly along the isobars will probably be regarded by many meteorologists as inaccurate. So it may be, and probably is, for any particular occasion, but the deviation is too small to be identified by direct observation and we have no adequate grounds for assuming that it is on one side more frequently than on the opposite. Hence, as a working principle the flow along isobars is the best general approximation which we can give.

west to east round the pole. It does not extend entirely to the equator, for in the equatorial region the current of the upper air is generally from the east up to great heights.

It must also be remembered that these are average conditions; the current may contain local circulations in whirls which are carried along with the current but the resultant motion is along the resultant isobars. This conclusion is borne out by a number of observations of the direction of motion of cloud in many parts of the world.

But to get at the explanation of our weather we want the distribution of pressure at the surface and we have so far only that at 4000 metres. That is certainly transmitted to the surface; but we must add the part which belongs to the lower layers; and here we must remember that for those levels, according to our maps, equatorial air is warm and polar air is cold and a column of warm air is lighter than one which is at the same pressure and is cold. When we take account of the lower strata, those below 4000 metres, we get an entirely different picture of the distribution of pressure, which is represented for summer and winter and for the northern and southern hemispheres by four maps (Figs. 28–31, plates xx–xxiii).

The law of relation of air-flow to the distribution of pressure still holds with the slight modification that owing to the friction of the surface there is a component across the isobars from high pressure to low. It is customary to regard the flow from high to low as the first step in circulation, but the first step in forming the atmospheric circulation was taken so long ago that we need not now repeat it.

So we have arrived at a representation of the average flow of air over the surface of the globe in winter and in summer which will enable us to explain many of the most important features of weather.

We see notable centres of high pressure round which the air circulates, some of these are permanent, such as those of the oceans which guide the trade-winds, the prevailing westerlies of the northern hemisphere and the roaring forties of the southern hemisphere.

Some of them are seasonal, namely those over the continent of Eurasia which give us the monsoons and over America which give us the weather of the American summer and winter. Some of them only last a few days but they give us a guide to our weather as when our winter cold flows from the great anticyclone of Asia over the cold European continent.

But these general features are only averages and only correspond with the daily weather for typical occasions. In the regions of the prevailing westerlies of the northern hemisphere and of the roaring forties of the southern latitudes there are local whirls whose life history has been a subject of discussion for fifty years. They pass over rapidly and give rapid changes of weather from fine to wet and back again which you will find mapped in our Daily Weather Reports and in the maps of some London daily newspapers.

Such is the outline; the weather of every day supplies the details and asks questions to which in time we shall be able to give an answer.

5. THE FIRST CHAPTER IN THE STORY OF THE WINDS

THE HALLEY LECTURE IN THE UNIVERSITY OF OXFORD, 28 MAY, 1918

1. INTRODUCTION

THE University of Oxford has no professor of Meteorology: nor is there any professorship of meteorology in any university of the British Empire. If you make inquiry, of those who should know, why that can be said of a subject which offers so many forms of exercise for the human mind as the study of weather and has so much interest, historical as well as practical for every human being, you will generally be told that meteorology has not arrived at the stage suitable for the attention of a professor, that it is not an exact science and that its votaries still spend their efforts mostly in accumulating facts.

I wish to maintain this afternoon the paradoxical thesis that if you look into the first chapter of the history of dynamical meteorology you will find that it is the dynamics and physics which have been inexact; and that the gradual accumulation of facts, which are exact, justifies us in asking for a new chapter based on a more rigorous dynamical and physical theory than that which is still generally accepted and one which would not be unworthy of the attention of a professor. This demand for a new chapter is offered as a respectful memorial of Edmund Halley, astronomer and physicist, who was the first to attempt a physical explanation of the origin of the trade-winds and monsoons in a paper published in the *Philosophical Transactions*, vol. xxvi (1686–7), pp. 153–168, when Halley was 30 years old, with the title: "An historical account of the trade-winds and monsoons observable in the seas between and near the tropics with an attempt to assign the physical cause of the said winds."

At that time contributors to philosophical transactions had no scruples about attributing the winds of the Atlantic to the exhalations of the seaweed to be found there, and Halley cut the first step to the true explanation of the general circulation of the atmosphere by introducing the principle of thermal convexion as a primary physical cause. It remains a primary physical cause still, though the method of its operation is much more difficult to make out than Halley or his successors supposed. The trade-winds, as everyone knows, are permanent currents of air near the line of the tropics from the north-east in the northern hemisphere and from the south-east in the southern hemisphere, while the monsoons are alternating winds in the East Indian Seas, about six months of the year from the north-east and about six months from the south-west (Fig. 32). Halley declined the assistance of the rotation of the earth as a dynamical agency. He accounted for the easterly component of the trade-winds by asserting that convexion followed the sun; but he based an explanation of the other component of the drift, namely that from the tropics to the equator, and the reversal of the monsoons, upon the convexion due to the greater heat of the equatorial region in the one case or of the land and sea areas respectively in the other. It was not until 1735 that the effect of the rotation of the earth was invoked by George Hadley[1], another of the Oxford fellows of the Royal Society, to explain the permanent direction of the trade-winds; and the theory took a form which after being rediscovered by John Dalton, Immanuel Kant and probably others found a place in every text-book of physical geography and which needs revision from the dynamical point of view.

2. Thermal Convexion

Halley drew from the principle of thermal convexion the idea that the natural basis of the explanation of the trade-winds and monsoons, the "cell," so to speak, of the atmospheric life, is a vertical circulation which follows directly from the visible thermal conditions at the surface and which any schoolboy is supposed to understand: there seemed, and to most people there seems still, no difficulty in supposing that where the surface temperature is highest the air rises continuously through its environment until it flows away at the top; just as it would do over a candle in the laboratory. The horizontal movement which constitutes the winds has simply to adjust itself to supply the vertical circulation. The same idea persisted from Halley's time to the present and

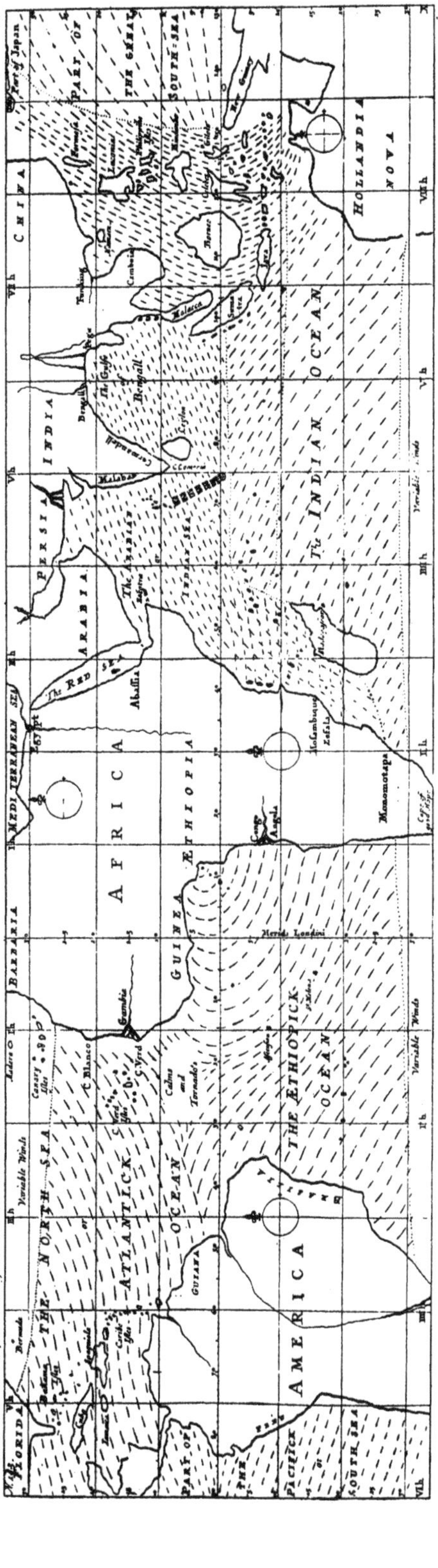

Carte des vents par Halley: Philos. Transactions 1688. Facsimile.

Fig. 32. Halley's Map of the Winds.

[1] *Phil. Trans.* vol. xxxix, pp. 58–62, 1735.

I wish to show, on the contrary, that the patient accumulation of exact observations has proved beyond dispute that it is the horizontal circulation over the earth's surface that is subject to a general law quite easily understood and verified, while the vertical circulation presents the most intricate difficulties, has never been actually verified and is not by any means fully understood even by the most learned among physicists or meteorologists.

To put the matter in a brief and bald way: according to the physical principles which Halley enunciated and which are now taught to everybody, you have only to identify the localities of maximum warmth and you thereby identify the positions of continuous rising currents of air which cause the winds; whereas according to the meteorological principles which are barely taught at all, you must look for the distribution of pressure and that alone in order to determine the character of the horizontal circulation of the winds. The vertical circulation is complex and indescribable in simple terms; the general picture of the horizontal circulation, duly verified by observation, even a schoolboy can easily carry in his head. Under the name of *Buys Ballot's law* the meteorological principle of the relation of wind to the distribution of pressure has long been known to be applicable in the regions north and south respectively of the belts of high pressure that surround the earth just outside the tropics, and the cause of the relation has been identified as being the rotation of the earth, under the influence of which, moving air changes its direction by $\sin \phi \times 30°$ per hour, to the right in the northern hemisphere, to the left in the southern hemisphere, unless it is prevented from doing so by the restraining force of pressure. In this expression ϕ is the latitude, and therefore $\sin \phi$, and the turn of the wind in an hour, diminish with the latitude; so the intertropical regions, where ϕ the latitude is very small, are regions in which the controlling effect of the rotation of the earth is correspondingly small, and Buys Ballot's law has less chance of asserting itself than it has further north on the one side or further south on the other. In fact, as a traveller crosses the line from north to south he passes from a region where Buys Ballot's law is operative in one sense, over the equator where it is not operative at all, to a region where it comes into operation again in the opposite sense.

It is because the regions of the trade-winds and monsoons are just the regions where the principle of the vertical circulation has the best chance and the principle of the horizontal circulation the worst chance that I think an examination of the two alternatives in those particular regions may be a fitting memorial of Halley's genius.

Halley's representation of the facts about the trade-winds and monsoons as given in the map in his paper was remarkably good from the observational point of view. He could not take the standpoint that I am taking to-day because he had not the necessary facts. He regarded the winds as currents of air from regions of low temperature finding regions of high temperature. The objection to doing so is that there are many currents of air that have no region of high temperature in front of them and low temperature behind

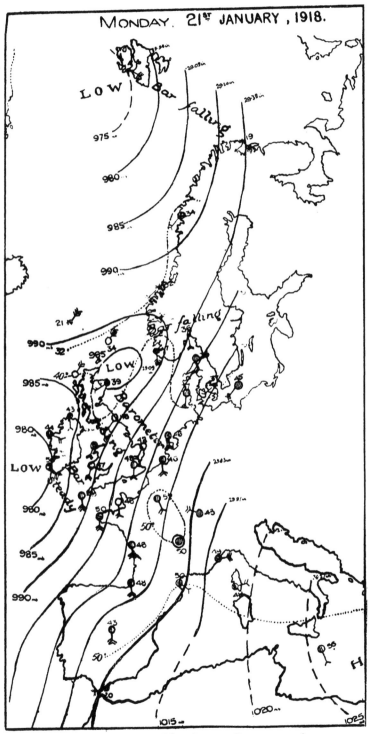

EXPLANATION

BAROMETER: Isobars are drawn for intervals of five millibars.

WIND: Direction is shown by arrows flying with the wind.

Force, on the scale 0–12, is indicated by the number of feathers.   Calm ○

TEMPERATURE: Given in degrees Fahrenheit. Isotherms shown by dotted lines.

WEATHER: Shown by the following symbols:

○   clear sky.

◐   sky ¼ clouded.

◐   sky ½ clouded.

◑   sky ¾ clouded.

◑   overcast sky.

●   rain falling.

✳   snow.

▲   hail.

≡   fog.

≡°   mist.

⊤   thunder.

⍅   thunderstorm.

Note that the wind guided by the isobars would travel from Madeira to Spitsbergen.

Fig. 33. Weather Map for 21st January, 1918.

them, and on the other hand there are hot regions and cold regions without currents to connect them. Let us regard the currents of air as circulating in the northern hemisphere round regions of high pressure on the one hand, the right hand, or round regions of low pressure on the other hand, the left hand. For the southern hemisphere you have only to exchange hands. The conditions are illustrated by Fig. 33, which indicates a flow from the south leading from Madeira to Spitsbergen.

3. The Equivalence of Pressure Distribution and Wind in the Upper Air

For some years I have thought that Buys Ballot's law is to be regarded as foreshadowing a still more important truth, as the manifestation at the surface, imperfect on account of surface friction, of a practically perfect correspondence between pressure and wind in the upper air and I have urged that meteorologists would do well to act upon the principle that in the free air the winds are tangential to the pressure-lines, the isobars, and there is a definite numerical relation between the separation of the isobars and the velocity of the wind, or at least that the divergence of the actual wind from the wind so defined and calculated is too small to be detected by any means at present at our disposal for drawing isobars or measuring winds. The time-scale of the operations which tend to cause deviation is so large that the course of the operations escapes observation.

Obviously such a principle entirely changes the outlook of meteorological investigation. I should like to give some justification for it. If you look at any meteorological chart of winds and pressures for the sea, or for land that is not mountainous, you find that Buys Ballot's law has hardly any exceptions. It is the one meteorological statement that you can always verify if you get a map. It smiles at you from every part of the map of the pressure and wind in the region of the British Isles. We have the advantage of dealing with a large number of stations on the sea coast in exposed situations not much above sea-level. Hence our winds and our isobars refer practically to the same level. Those who deal with the meteorology of large continents are not so fortunate. The wind does not lie exactly along the isobars, it is true, but it always pays attention to the isobars and generally (with due regard to diurnal variation and other considerations) it pays attention also to their distance apart. And here we must remember that the winds are measured at the surface. They are in the worst possible position for arranging themselves according to any principle on account of the obstruction to free motion which the surface offers, and yet they always show that they bear it in mind, and we know from actual observation that up above the surface the agreement between the actual wind and the calculation from the surface isobars gets still better. It is impossible to carry out a strict numerical verification of the law of relation of wind to pressure because in order to do so we must get both for the same level and until aircraft comes to our aid, we cannot, either by observation or calculation, get accurate measurements of the pressure-distribution at a level where the winds are free from the effects of the surface. And, moreover, so little pressure-difference means so much wind—a tenth of a millibar in a mile means a gale—that the accuracy necessary for direct comparison is quite at the limit of the power of our instruments. There is no doubt that the balance between wind and pressure is the condition to which the whole atmosphere tends when it is free from disturbing forces, and any new disturbing cause can only operate through the medium of the atmosphere. When people think of the causes of winds they generally imagine to themselves a calm atmosphere out of which winds are to be made by setting up a finite difference of pressure; it is not rest or calm that we ought to look upon as the normal condition of the atmosphere upon which our available causes must act, but an adjusted or nearly adjusted balance of wind and pressure, and the smallest disturbance immediately calls out the forces for restoring the adjustment. The effect of viscosity upon the relative motion of the free air was shown by Helmholtz to be extremely small. There are many things that we can explain and hardly any that we cannot explain by supposing that the surface-winds simply show the calculated effect spoiled to some extent by surface retardation; and, in fact, the most reasonable thing to do, at any rate for our latitude, is to accept the relation between pressure and wind as a principle, and work from it to conclusions and examine the results until we find it can carry us no further.

The region between the tropics, as already mentioned, is not a favourable place for applying the principle because the rotation of the earth upon which the adjustment depends is so little felt. Let us see then what we can make out of it for that unfavourable region.

4. MODERN MAPS OF THE INTERTROPICAL REGION

First let me exhibit the maps in the possession of the Meteorological Office of the winds in the intertropical belt round the globe which are the most modern equivalent of Halley's map (Fig. 34)[1]. They show the regions where the well-known features of the trade-winds and monsoons are most conspicuous. The original maps are in four sections, the Atlantic Ocean, the Indian Ocean, the West Pacific Ocean and the East Pacific Ocean, and they represent the information about wind and pressure for the months of June and December collected in a long period of years, partly in this country, since 1854, partly in Germany, and partly in the United States. The information for the Pacific Ocean is still insufficient for really satisfactory maps, there is less objection to the maps of the Atlantic and Indian Oceans on that score, and as those represent the regions of the trade-winds and monsoons as generally understood I shall ask your attention principally to them. What I have to say about them depends upon detail, but to enter into all the details represented on the maps would take too long, so I will restrict myself to a few. But first let me explain that in current literature the words "trade-wind" and "monsoon" have come to be used loosely and to have a more general significance than is associ-

[1] *Barometer Manual for the Use of Seamen.* M.O. Publication, 61.

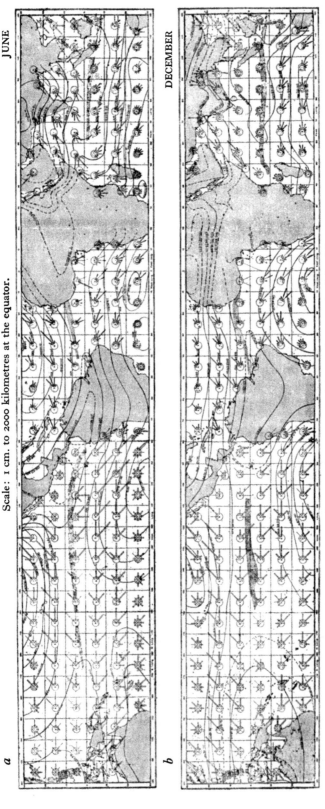

Fig. 34. Charts of Normal Pressure, and Wind-roses in the Intertropical Belt.

Scale: 1 cm. to 2000 kilometres at the equator.

a JUNE

b DECEMBER

Pressure. Isobars for even millibars (1010 to 1020 between the equator and the tropics over the Atlantic).

Doldrums. Shaded areas show the "doldrums" of the equatorial seas.

Wind-roses. For each 10° square; representing, by the lengths of the arrows, the frequency of winds from each of the sixteen even points of the compass, viz. 0°, 22½°, 45° etc. from true North. The scale for the arrows is approximately 1 cm. to 100 per cent. of frequency.

For the South Atlantic and Indian Oceans. The length of the single fine line from the point of the arrow within the circle represents the number of light winds (less than 5·5 m./sec.) as a percentage of the whole number of observations including calms, the double line moderate winds (5·5 to 17 m./sec.) and the "blocked in" outer end (if any) represents gales (greater than 17 m./sec.). The figure in the circle gives the percentage of calms.

For other oceans the mean force of the winds from the several directions is represented. A single line represents a mean force of light winds and a double line a mean force of moderate wind. Where there is not space enough for an arrow of full length the lines are broken and the percentage given in figures in the break. For details see the *Barometer Manual for the Use of Seamen*, Plates xxxvi–xliii, from which these charts are compiled.

ated with them in the ordinary explanation of their origin. In the northern hemisphere, for example, we have in theory a north-east trade-wind and a north-east or south-west monsoon, but in practice the trade-wind in the West Indies may blow from the east or south-east and still be known as the north-east trade, and, in the eastern seas, winds with a seasonal alternation of direction are monsoons whatever may be the specific directions between which they alternate. In fact the characteristics of the trade-winds are really taken, for the north-east, from the regions some way off the north-west coast of Africa between 30° N. lat. and 10° N., and for the south-east, from the regions between 30° S. and 10° S., and again some way off the West African coast. Those are the regions where for centuries ships found and utilised the trade-winds, and any winds that might be regarded for the purposes of navigation as a continuation of the winds of those regions were still called trade-winds. Similarly, the characteristics of the monsoons are taken from the Indian Ocean and the Bay of Bengal and as the area of experience extended the new facts were joined on to the old, retaining the same terminology.

5. The Winds of the South Atlantic in December

Let us look particularly at the map of the South Atlantic for December and note the region of highest pressure between 0° and 10° W. long., with oval isobars round it extending from 30° S. to within 5° of the equator, and note that the regions of characteristic south-east trade form the northern and eastern part of this area: that the main feature of the whole is a circulation round the central area of high pressure very steady on the eastern side but variable on the western side. The variability may be attributed to alterations, perhaps from day to day or perhaps in different years, of the shape of the isobars on that side. In the Bight of Africa the winds blow towards the African shore and there is a corresponding break in the regularity of the isobars.

The air of the north-east trade turns westward between the equator and 10° N., the south-easterly seems more determined to reach the equator. Between the two is the region of calms and variable airs with heavy rain and thunder-showers, called the doldrums of the equator, which on the convexion theory is supposed to be the aim and object, the final destination, of the air that flows from both sides of the equator, the controlling cause of the whole motion; but the doldrums are not regions of continuous rain, they have their intervals of fair weather and blue sky. The principal destination of the wind that flows towards the equator from either side seems to be to flow along the equatorial belt and up the valley of the Amazon, and the flow of the Amazon river represents the collective result. On the map one wind-rose is shown centred in the doldrum region: the winds are light, they are not the final rush of mighty currents that control the circulation of the air over the globe, they are much more aptly described as incidents of a region where the rotation of the earth has no controlling influence. The key to the atmospheric situation is not so much the warmth of the equatorial belt as the regions of high pressure north and south of the equator, round which the air circulates with some slight drift of the surface air towards the equator.

6. The Monsoons

Next let us look at the maps of the Indian Ocean and consider the course of the monsoons which are characteristic of that Ocean. According to the convexion theory the S.W. monsoon in June is the flow of air from cold sea to the heated land of the Asiatic continent. In the map you will see it as the flow of air guided and controlled by isobars which show the distribution of pressure north of the equator. And as one looks at the wind-roses marking a trade-wind south of the line, it certainly looks as if the south-east trade crossed the line between 40° and 70° E. longitude and passing round the end of a line of doldrums along the equator from 60° E. to Sumatra formed the feeding current of the south-west monsoon. That is a view of the origin of the south-west monsoon which is generally accepted without demur. But please note that the pressure-lines and winds are given for the sea only, and that circumstance tends to favour the appearance of self-contained circulation over the ocean. The Meteorological Office, however, owes to Colonel H. G. Lyons a new set of maps of the distribution of pressure over the Mediterranean basin and Africa[1] which throw a new light upon the origin of the south-west monsoon. His maps for the summer months show a steep gradient of pressure for northerly winds over Egypt, the Red Sea and Arabia, and the pressure-lines which represent the gradient join those which are shown in our maps over the Indian Ocean and form the guiding influence of a great circulation round an area of low pressure north of the Persian Gulf. The winds are in admirable agreement with the pressure-lines and thus we may trace at least the greater part of the south-west monsoon to a great current of air which flows from the north over the western Mediterranean, the Nile valley, the Red Sea and Arabia, turns round very much in the same way as the Red Sea does itself and passes westward across the Indian Ocean as the south-west monsoon.

Thus the south-west monsoon also obeys with singular fidelity the distribution of pressure. It is not so attentive to the distribution of heat, for if there is a hot place on earth in June or July where should it be if not in Arabia or the Sahara? Why should the wind neglect those attractive centres and seek another locality if the temperature is the direct controlling cause? So in like manner the north-east monsoon is a circulation round the high pressure formed in the winter season over the great Asian continent.

7. Isobars crossing the Equator

I am still uncertain what to do with the winds that appear to cross the equator to form, or to add to, the south-west monsoon. They also seem to

[1] *Monthly Meteorological Charts of the Mediterranean Basin*, M.O. 224 (2nd edition), 1917 and *Q. J. Roy. Met. Soc.* vol. XLIII, p. 113 *et seq.*, 1917.

follow isobars that lead to and cross the equator. I always find my pen sticks in the paper and refuses to move when I try to draw an isobar across the equator. On the north side of the equator the low pressure is on the left of the wind and the high pressure on the right; south of the equator the reverse is the case, so it seems that in passing from the south with the high pressure on the left across the line to the north where the high pressure is on the right, we pass a region where there is no relation at all between wind and pressure. So an isobar loses all meaning as regards wind at the equator and there ought at least to be some way of indicating that on the map.

The best way would be for the isobar to double back from the equatorial region so that the pressure systems of the two hemispheres would keep themselves distinct. It seems doubtful if they could do otherwise in the case of a synchronous chart, but whether by taking means the arithmetic may give something which nature in reality abhors I cannot say. On the map in question the isobars are drawn according to pressure values compiled from the collection of barometer readings for "squares," of one degree, two degrees, or five degrees, and a more detailed study of the distribution of pressure in the belt of ten degrees on either side of the equator would probably give us the material necessary for settling the present question. What is certain is that isobars crossing the equator are more frequent in proportion as our knowledge is less, and I hazard the suggestion that they tend to disappear on closer investigation. There are none in the Atlantic, the equatorial region of which has been most carefully studied: they are of little importance in the maps of the Indian Ocean which are based on a large number of observations; but in the Western Pacific for which observations are comparatively scarce they seem to form the rule and not the exception, and there the relation of isobars to winds is least marked. Further investigation is much needed. In the meantime let us notice that the convective theory of the trade-winds is equally applicable to any part of the earth's surface but really we only find the trade-winds developed according to rule in certain special localities. Instead of considering the scheme of flow as a general advance all round the earth towards a line of calms and variable winds at the "heat-equator" of the earth, what we really have is a pair of belts of high pressure along the tropics north and south of the equator with highest pressures at 30° N. or S.; they are permanent over the oceans and seasonal over the land and they are broken up partly by the continents or by other causes into large but isolated areas of high pressure round which the winds circulate. On the eastern sides of these areas we have the trade-winds in their most fully developed form and there they are most persistent in direction. On the western sides the winds find their way back again from the equatorial region past the tropics but with far less regularity than the trades. That the return flow from the equatorial region is not simply a question of continents may be illustrated by a comparison of the winds at St Helena, in 16° S., 6° E., and at Suva, Fiji, in 18° S., 178° E. The St Helena winds are quite persistently from the S.E., the Suva winds, on the other hand, congregate mainly round N.E. (see p. 26).

Note. October, 1921. On reading this passage again I think I have now clearer light upon the peculiar relations of the winds to the isobars in the equatorial part of the Indian Ocean during the S.W. monsoon and on the Guinea coast of Africa, and the crossing of the equator by the air which apparently flows along the surface to supply the monsoon and the Nigerian rainfall. The usual conditions of pressure in the equatorial region as exhibited over the Atlantic and Pacific Oceans are represented by an equatorial belt of low pressure between the two belts of high pressure on the tropics. Such a distribution favours easterly winds in those regions in which the winds are dependent on the pressure distribution. Thus the equatorial belt of air (which is not subject to any such guidance) lies between the two currents of east wind and in the absence of any other controlling influence may be carried along to the westward by the east winds on its flanks, and so form with them a belt of heterogeneous east wind between the tropics.

Reference to the isobars of the maps of Fig. 34 will confirm this view and also show that in the Indian Ocean, in the summer, the equatorial trough of low pressure is merged in the general area of low pressure belonging to the monsoon-circulation, and there is a *continuous slope of pressure* from the high pressure region of 35° S. on either side of Africa across the equator to the low pressure regions of the monsoon area or its extension westward over Western Africa.

With a continuous slope of pressure crossing the equator we have the peculiar conditions that the flow in the area north of the equator ought to be from west to east, and south of the equator from east to west, and so in fact it is. Between the two currents is a line of equatorial doldrums and the feeding of the monsoon by crossing the equator appears to take place in the western part of this doldrum region round which the winds circulate, so far as concerns the westward moving air on the south side, as though it were an elongated area of low pressure in less peculiar parts of the world. The region is apparently marked by a separation of the isobars indicating a flattening of the slope of pressure there and the central belt of the right-handed circulation exhibits in this case doldrum-weather, not the weather of an anticyclone.

What compensation the northern hemisphere makes for the air which is thus derived from the southern hemisphere it is not easy to say. The conditions are apparently repeated on a very mild scale off the southern coast of Mexico and Central America. There is no corresponding area where there is a continuous slope of pressure from north to south across the equator where the southern hemisphere borrows again from the north. But the difficulty is not a serious one because it could be met by a very small drift southward over a large front, or by some adjustment in the upper air.

The ideas which are thus formulated for the representation of the inter-tropical region in July are given in Fig. 35 which shows the crossing of the equator as well as the movement along it in opposite directions on the two sides.

8. NUMERICAL CALCULATION OF THE VELOCITY OF THE TRADE-WINDS

Next let us approach the subject numerically. Comparing the actual velocities of the wind in the north-east trade and the south-east trade let me quote some figures for each taken from another official publication[1]. They are compiled partly from observations of pressure and wind obtained during about 40 years from ships that travelled in those regions, and partly from an anemometer at a height of 2000 feet in the heart of the trade at St Helena in latitude 16° S. The values are given in miles per hour.

	Jan.	Feb.	Mar.	Apl.	May	June	July	Aug.	Sept.	Oct.	Nov.	Dec.	Year
N.E. Trade	10	11	11	12	11	10	9	7	8	6	8	10	9·4
S.E. Trade	14	13	13	12	11	12	12	15	17	15	16	15	13·8
Mean	12	12	12	12	11	11	10·5	11	12·5	10·5	12	12·5	11·6

The uniformity of the mean values is very remarkable. And, thirdly, let us compare the south-east trade as recorded by the anemometer at St Helena, very much in the free air, with the values of the wind velocity calculated from the distribution of pressure in the neighbourhood.

	Jan.	Feb.	Mar.	Apl.	May	June	July	Aug.	Sept.	Oct.	Nov.	Dec.	Year
Calculated	16	15	16	15	11	12	14	15	15	15	17	16	14·8
Wind of St Helena	16	15	15	14	13	11·4	14	18	20	18	19	17	16·2

The agreement is not perfect but it is sufficiently close to show that the anemometer at St Helena keeps the distribution of pressure in mind.

As regards the surface, therefore, we may fairly conclude that except for a narrow strip along the line of the equator the winds are generally in accord with the distribution of pressure. There are parts of the circulation in which the air moves towards the region of higher temperature, but when it has to choose between that direction and obeying the pressure it obeys the pressure. When the physical hypothesis is in conflict with the meteorological presentation the meteorology wins. And there is really no reason to be dissatisfied with the result, for the dynamical theory of the relation of the wind to the distribution of pressure accounts for the facts in a satisfactory manner, explaining the deviation from the pressure-lines as the effect of the friction with the surface. The theory explains that if left to itself air, or anything else endowed with a free motion in any direction over the earth's surface, will turn its direction through an angle $\sin \phi \times 30°$ per hour (where ϕ is the latitude) in consequence of the rotation and rotundity of the earth. To keep it from turning it must be pushed, that is, a suitable distribution of pressure is required. It is really difficult to understand why people have been so long content with Hadley's setting out of the effect of the rotation of the earth. He wanted to explain a deviation from north to north-east, that is a rotation of 45°. We know now that on the hypothesis of flowing away to the southward, if free to turn, the air would change its direction by $\sin \phi \times 30°$ per

---

[1] M.O. 203, *The Trade Winds of the Atlantic*.

Fig. 35. Normal Pressure and Winds in the Intertropical Region.

Scale: 1 cm. to 2000 kilometres at the equator.

JULY

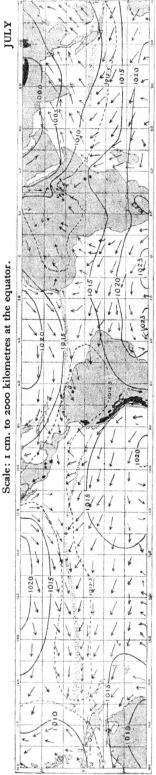

Pressure. The distribution is shown by isobars, generally for steps of 5 millibars from 1000 to 1020 (1023 for the South Atlantic). On account of the points of interest which it discloses the isobar for 1012·5 mb. is also shown. Doldrums. The probable position in the several oceans is shown by shading. Wind. The prevalent direction only is shown, by arrows which fly with the wind.

The information has been derived from the following sources: Barometer Manual for the Use of Seamen, M.O. 61 (1919); Monthly Meteorological Charts for the East Indian Seas, M.O. 181; Monthly Meteorological Charts of the North Atlantic Ocean, M.O. 149; Wind Charts of the South Atlantic and of the Coastal Regions of South America, M.O. 168 and 159; Pilot Charts of the North Pacific and South Pacific Oceans, U.S. Navy Hydrographic Office. Pressure maps for Africa, Colonel H. G. Lyons, F.R.S., *Q.J.R. Met. Soc.* vol. XLIII, p. 113, 1917.   The land areas above the level of 4000 metres are blacked.

The reader's attention is drawn particularly to the continuous slope of pressure over the Indian Ocean across the equator from 1020 mb. at the tropic of Capricorn to 1000 mb. at the tropic of Cancer, with opposite flow of air on either side of the doldrums of the equator, connected by a flow across the equator between Africa and the region of doldrums thus forming a quasi-clockwise circulation in a continuous slope of pressure of which the portion north of the equator constitutes part of the S.W. monsoon. The system extends over Africa and shows a continuous slope from 1023 mb. in the South Atlantic to 1005 mb. along the Red Sea associated in like manner with a diversion of the S.E. trade across the equator, and from sea to land at the Gold Coast, forming another quasi-clockwise circulation round the head waters of the Congo and the Nile.

Note may be taken of the flattening of the slope of pressure and probable disappearance of pressure-gradient in the region where the wind does not cross the equator. The contrast of conditions for crossing with those for not crossing is very clear in the map for June, Fig. 34 *a*.

In the western hemisphere there appears to be a continuous slope from 1020 mb. in the S.E. trade-wind of the Pacific across the equator to 1010 over the Gulf of California. The winds in the neighbourhood are irregular but no general system is decipherable.

Corresponding maps of the circulation of the upper air over the intertropical regions as indicated by cirrus clouds have been compiled by Professor W. van Bemmelen. See *Nature*, 14 Jan. 1922.

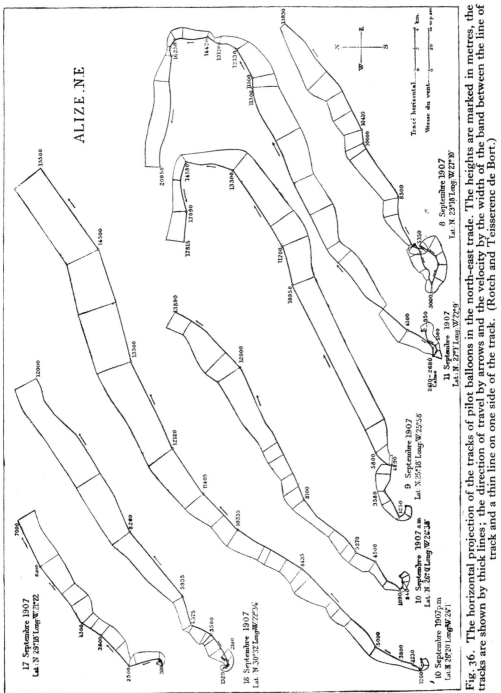

Fig. 36. The horizontal projection of the tracks of pilot balloons in the north-east trade. The heights are marked in metres, the tracks are shown by thick lines; the direction of travel by arrows and the velocity by the width of the band between the line of track and a thin line on one side of the track. (Rotch and Teisserenc de Bort.)

hour; a turn of 45° would take 3 hours near latitude 30°. But the air travels far more than 1000 miles occupying about 10 days for the journey; why then does it stop its rotation at 45°? Only because it is coerced by the neighbouring air. So the coercion of the neighbouring air must be the real controlling influence in any case.

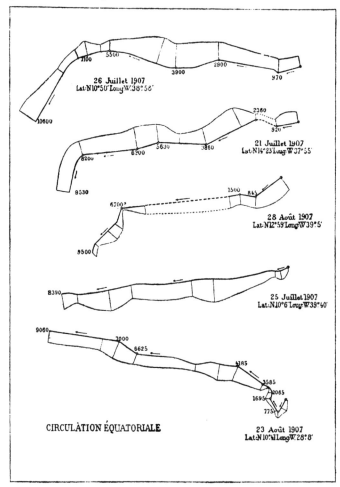

Fig. 37. For explanation see Fig. 36.

9. THE UPPER AIR

If this be a true representation of the facts for the surface the same reasoning must apply to the upper layers, say at a kilometre above the surface, and at successive kilometres above that, with this modification—that the frictional effects of the surface no longer interfere with the balance of pressure and velocity, so if we can trace a relation between the flow of air and the isobars at the surface we must insist upon an even stricter relation of the same kind higher up.

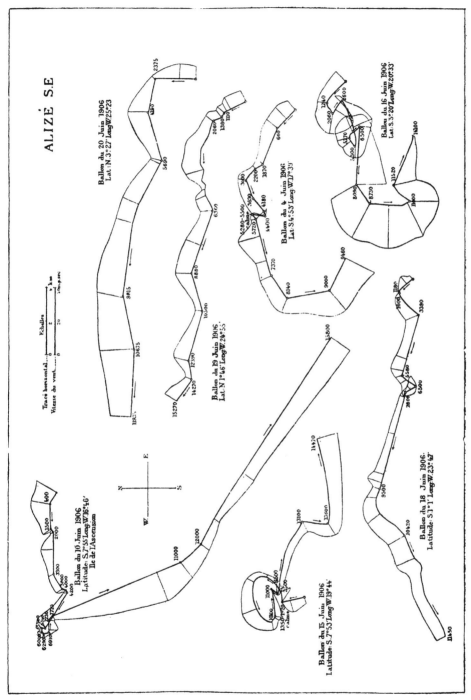

Fig. 38. For explanation see Fig. 36

That proposition compels us to face considerable changes in the distribution of pressure at higher levels from that which we see on our maps at the surface. The winds of the trade region have been explored by the Yacht "Otaria," chartered for the purpose by two enthusiastic meteorologists whose loss we still deplore, M. L. Teisserenc de Bort, of Trappes, France, and Mr A. L. Rotch, of Blue Hill, Boston, U.S.A., and the results are shown in the illustrations (Figs. 36, 37 and 38) taken from the published account[1]. They show the horizontal projections of the tracks of balloons sent up either from ship or shore and watched from below. The observations refer to the north-east trade, the equatorial region, the south-east trade and Ascension, so that we have here a new exploration of the trade-winds[2] that is supplementary to Halley's. The story will not be complete until we can compare the diagrams with the distribution of pressure at the same levels.

10. Pressure in the Upper Air

We have not enough information about pressure in the upper regions over the sea to draw mean isobars at different levels, but we have something to go upon for North Africa and the Mediterranean shores, where there are high level as well as low level stations. The data have been put together by Colonel Lyons in the form of maps for sea-level, the 2000-metre level and the 3000-metre level[3]. The data are not numerous, but they enable us to get an idea of the conditions of pressure and wind at those levels which, so far as they go, show no tendency to contradict our principle that the winds adjust themselves to the distribution of pressure. The July winds at Teneriffe and St Vincent are found to belong to a narrow belt of low pressure. For a still higher level, that of 4 kilometres, we have some maps computed from the surface pressures and temperatures by M. Teisserenc de Bort (Figs. 24–27, plates XIV–XVII), and with those can be associated the motion of clouds as ascertained by observations from the surface. These have been carefully co-ordinated by Professor Hildebrandsson of Upsala[4] and they agree remarkably well with the distribution of pressure.

We must not give too much weight to the details of the lines in the intertropical region because the temperature data of the upper air are not sufficient for the computation; but the general result is worthy of attention because it leads up to the question of the counter-trades, which have been known for a long time to prevail above the trade-winds and have always been regarded as confirming the convexion theory of the trade-winds by showing the return current required to complete the simple vertical circulation that every schoolboy can understand. The tracks of the balloons show a good deal of variation in the direction of the upper wind. Counter-trades from the south-west over the north-east trade and from the north-west over the south-east trade require areas of low pressure over the anticyclones of the north and south

---

[1] *Travaux Scientifiques de l'Observatoire de Trappes*, tome IV, 1909.

[2] In the original publication a map showing the track of the ship, diagrams of the projections of the tracks of pilot balloons sent up from the Azores, Madeira and Teneriffe are also included.

[3] *Q.J. Roy. Met. Soc.*, vol. XLIII, p. 113.   [4] *Ibid.*, vol. XXX, p. 317, 1904.

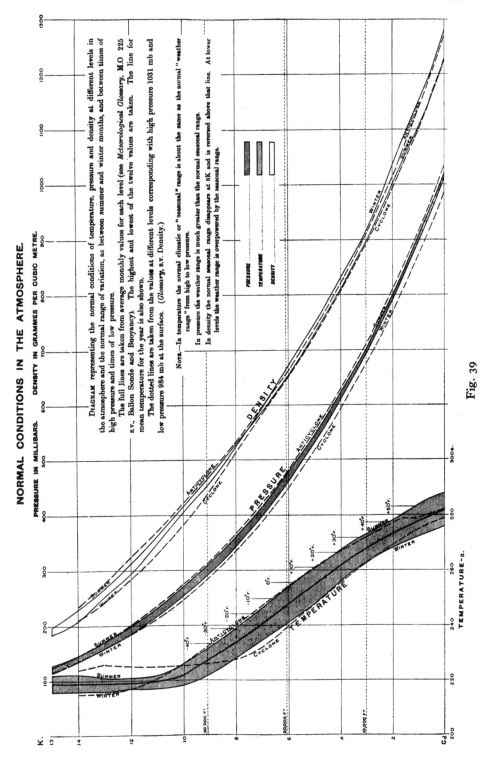

NORMAL CONDITIONS IN THE ATMOSPHERE.

PRESSURE IN MILLIBARS.   DENSITY IN GRAMMES PER CUBIC METRE.

Diagram representing the normal conditions of temperature, pressure and density at different levels in the atmosphere and the normal range of variation, as between summer and winter months, and between times of high pressure and times of low pressure.

The full lines are taken from average monthly values for each level (see *Meteorological Glossary*, M.O 225 s.v., Ballon Sonde and Buoyancy). The highest and lowest of the twelve values are taken. The line for mean temperature for the year is also shown.

The dotted lines are taken from the values at different levels corresponding with high pressure 1031 mb and low pressure 984 mb at the surface. (*Glossary*, s.v. Density.)

Note.—In temperature the normal climatic or "seasonal" range is about the same as the normal "weather range" from high to low pressure.

In pressure the weather range is much greater than the normal seasonal range.

In density the normal seasonal range disappears at 8K and is reversed above that line. At lower levels the weather range is overpowered by the seasonal range.

Fig. 39

Atlantic. The isobars of Teisserenc de Bort's maps suggest that the areas of low pressure in the upper air are not circular or elliptic areas like the anti-cyclones at the surface, but bends in the lower members of the great system of westward isobars which produce the same result so far as the regions of the counter-trades are concerned (see also Figs. 75 to 78).

11. The Vertical Circulation: Environment

So much for the horizontal circulation. Now let us go back to the vertical circulation and look more carefully into the details.

The conventional assumption is that wherever the air is warmer than its horizontal environment it will ascend and a continuous flow of air will be maintained making the up-current of a simple vertical circulation that controls or induces the surface winds. Let us be a little more exact in our physics and remember that in the atmosphere going upward means reduction of pressure and that, for dynamical reasons, reduction of pressure means a corresponding reduction of temperature; going downward means, *vice versâ*, an increase of pressure; and an increase of pressure means an increase of temperature.

The air begins to rise because it is warmer than its environment and as it rises its temperature falls automatically. Will it then still be warmer than its environment and ascend still further, or will a short upward journey suffice to bring it into equilibrium with its environment at a very moderate height? That depends of course upon its environment in the upper layers of which Halley could know nothing but of which we at least know something.

The next diagram (Fig. 39) that I have to show represents the results of ten years of W. H. Dines's work and gives the normal condition of the environment in Southern England. You see there the average temperature, pressure and density at all heights up to 15 kilometres in summer and winter, in average cyclonic conditions and anticyclonic conditions. The next diagram (Fig. 40) shows what happens to air containing different quantities of water-vapour as its pressure is reduced by elevation. Now, in order to decide how far a particular specimen of air will rise we have to bring the corresponding parts of these two diagrams together. The lapse-rate for dry air is indicated by the full lines at the bottom of

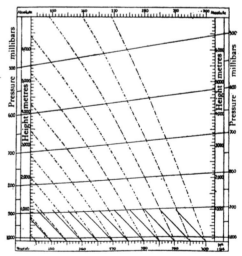

Fig. 40. Adiabatics for dry and saturated air. Neuhoff's diagram.

Fig. 40. It is quite evident that with a normal environment the air will hardly rise at all. The only way for the air to rise is for it to build up what we

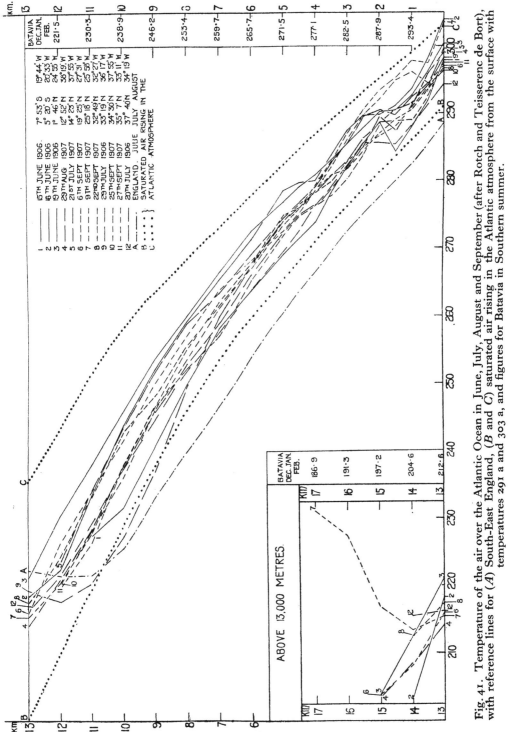

Fig. 41. Temperature of the air over the Atlantic Ocean in June, July, August and September (after Rotch and Teisserenc de Bort), with reference lines for (*A*) South-East England, (*B* and *C*) saturated air rising in the Atlantic atmosphere from the surface with temperatures 291 a and 303 a, and figures for Batavia in Southern summer.

call an isentropic layer and then any warming at the bottom of the layer will carry the warm mass to the top. We have the general proposition that if the slope is that of the adiabatic line ascent of warmed air is possible, but when the slope on the diagram is steeper than the adiabatic line the ascent is speedily arrested. Near the surface ascent is very unlikely because the lapse of temperature in the environment is relatively slight. When condensation begins the slopes of the two diagrams are nearly identical and ascent is much more likely.

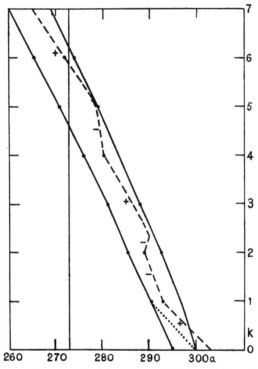

Fig. 42. Ascent of 16 June, 1906. Equatorial region. Lat. S. 5° 20'. Long. W. 20° 35'. Teisserenc de Bort.

The chain line shows the observed lapse of temperature with height. The full lines show the lapse in the adiabatics of saturated air, the dotted line shows the same for dry air.

Hence we may conclude that the regions where air ascends continuously are not necessarily the regions where the surface air is most heated, but where the lapse-rate in the air surrounding it is greater than that of the adiabatic.

Now let us look at some actual examples of the condition of the trade-wind region taken from M. Teisserenc de Bort's *Memoir* (Figs. 41 and 42). It will be seen to have some striking characteristics, in some cases ascent would be easy, in others difficult. For one of the ascents (in Fig. 42) I have marked with a + the places where ascent would be certain for saturated air, and with a − the places where ascent would be arrested.

You will agree that the climb of the surface air into the upper regions is a very tedious process, but it is possible: the return journey is hardly even that.

It is difficult to get air up, it is almost impossible to get it down again. If it starts in an atmosphere with the adiabatic lapse-rate for saturated air it will soon get warmer than its environment because its own lapse-rate is that for dry air. If it were mechanically pulled down from 10,000 metres it would be hotter than anything ever known at the surface. Let me show you a final diagram that represents the facts in this respect, an entropy-temperature diagram for dry and saturated air[1].

For a long time physicists have allowed meteorologists to think that anti-cyclones were made up of descending air, and they have therefore come to regard them as the localities where compensation was to be found for the ascent of air that is indicated by the rainfall of cyclonic depressions. They are generally large areas and it has been customary to suppose that they represented regions where air comes down from the higher reaches of the atmosphere to the surface. Suppose that were true and that the distribution of temperature in the upper air of an anticyclone were that represented by the continuous descent of a homogeneous supply of air. The result would evidently be to provide an environment which would be ideal for the prolonged ascent of air warmed at the surface. The slightest amount of warming at the surface would mean that the warmed air must necessarily go to the top—there could be no stopping place at all on the way up. They would be ideal places for the manufacture of local showers of explosive violence; instead of being, as we actually know them, regions of remarkable atmospheric stability, they would be just the opposite. Instability and sudden changes of weather would be their characteristic. The lower latitudes would be doldrums. An anti-cyclone is certainly not a channel by which the upper air "plunges downward" from heaven to earth.

12. SUMMARY

Let me review the course over which I have endeavoured to lead you this afternoon. I began with the very easy physics that Halley propounded, and which, if we trusted to it, would land us in extremely complicated meteorology, ascending currents wherever the surface air is warmer than its environment compensated by descending air of which no one knows the history; and I have substituted for it very simple meteorology, the most rudimentary principles of universal application combined with physics and dynamics, grimly real but so difficult as to terrify even the bravest physicist or meteorologist and to make him proceed with the most respectful caution. If there had been a professor of meteorology in Oxford what I have said would have been common knowledge for all of you, it would have come to you in the daily round like the milk or the morning paper, and I could have spent the fleeting hour over some of the interesting developments which perforce I have left untouched.

[1] This diagram in more developed form is given on p. 118.

And the story which a professor of meteorology would tell you is indeed a romance, call it the mating of pressure and wind, which might well replace some of the stories of the loves of Jack and Jill, in high life or in low life, that occupy so many of our library shelves to-day. It has all the elements of romance, there is love at first sight with which begins a helter-skelter wind across the isobars, but the relentless ways of the world, that inexorable rotation of the earth, makes the rash course miss its mark and the settled march of matrimony begins with high pressure on the right and low pressure on the left. Even then the course of true love does not run quite smoothly; the rubs of the hard earth or the troubled sea make deviations for which no remedy is to be found on the solid earth, but in the heaven above the mating is complete, pressure and wind live together happily, or very nearly so, ever afterwards. Even in those ethereal regions there may be occasional instability, a few storms of rain, perhaps snow or hail, or even thunder and lightning, but the Latin grammar tells us about them "amantium irae amoris redingratio est," they are merely temporary disturbances of that serene true course in which pressure always adjusts itself to wind and wind to pressure. To the story of that perpetual state of balanced happiness Halley contributed the introductory chapter, dealing with the period of love at first sight. A professor of meteorology could write for you another chapter, in which the earth's rotation has tempered the spasmodic passion and made possible the endurance of a permanent union. I have not given you the chapter; it is not for me, a mere official, to do so; but I hope I have shown you that the material is there, needing only the touch of professorial eloquence to make out of it for you (and all the world besides) the true story of the wind and the pressure.

6. THE DROUGHT OF 1921

REPRINTED BY PERMISSION FROM *The Times* OF 15 OCTOBER, 1921

THE weather of the British Isles in the year which has now run through three of its four quarters has been remarkable for its drought, its abundant sunshine and its exceptional warmth. It is the more remarkable because the exceptional dryness, sunshine and warmth have been continued half way through the month which a great artist pictured as "chill October," and which on the average of a long series of years is the wettest month of the year in the south-east of England. Our expectation of rain in October is 88 mm. or 3·46 inches as compared with 52 mm. or just over 2 inches in March. This year we were short of rain for March and it seems probable that we may be still shorter for October. The drought has been so exceptional and prolonged in this particular region that we have approached the conditions of the semi-arid districts of the world, where the normal annual rainfall is between 10 inches, or 250 mm., the irreducible minimum for crops, and the more or less comfortable 20 inches. In such semi-arid districts the variation from year to year attributable to ordinary meteorological causes makes all the difference between famine and plenty. Such regions are the famine regions of the world. Famine regions are not the desert regions, where it is known that crops require irrigation; they may border on deserts, or they may be themselves regions of low rainfall; and for the droughty years they become deserts incapable of supporting crops or stock. Such areas are to be found notably in India, on the border of the Australian arid region and in a peculiar form in the Ceara district of Brazil which presents "O secular problema do nordeste" of that country. The western provinces of Canada are near the line, so is Russia in Europe, and this year has provided an appalling illustration of what the line means. All Europe has complained of drought, but meanwhile I learn from the Director of Blue Hill Observatory, Massachusetts, that "July, 1921, was not only the wettest July since observations began at Blue Hill, thirty-six years ago, but the wettest month of any name." 265 mm. of rain were recorded against an average of 100 mm. "With the exception of August, 1826, the month had the heaviest rainfall in a period of 100 years."

THE DROUGHT IN ENGLAND, SCOTLAND AND IRELAND

Such are the vagaries which the revolving suns produce. Not even over the British Isles has the drought been equally severe. The Meteorological Office keeps the public memory of the weather for us, and from the files of the Weekly Weather Report we extract the following figures about the rainfall in England South-East (the counties south of the Thames as far west as Dorset), Ireland South, and Scotland, north of the Caledonian Canal, including the

Islands. They give the normal fall of rain for each week, the actual fall, and the aggregate excess or defect to date, and therefore show the progressive course of the drought from the beginning of the year. The figures are given in millimetres because they fall into convenient numbers for comparison, and that is the only purpose for which they are required, but anyone who prefers to work in inches, tenths and hundredths, can convert the millimetres into hundredths of an inch by multiplying by four.

WEEKLY RAINFALL, 1921

Dates and numbers of the weeks		England S.E.			Ireland S.			Scotland N.		
		Normal	Actual	Aggregate excess (+) or defect (−)	Normal	Actual	Aggregate excess (+) or defect (−)	Normal	Actual	Aggregate excess (+) or defect (−)
Jan. 2– 8	1	14	18	+ 4	25	28	+ 3	29	50	+ 21
9–15	2	12	23	+ 15	23	29	+ 9	29	22	+ 14
16–22	3	12	6	+ 9	20	10	− 1	34	90	+ 70
23–29	4	12	3	0	20	19	−−	40	72	+102
30–Feb. 5	5	13	9	− 4	20	33	+ 11	35	6	+ 73
Feb. 6–12	6	14	2	− 16	23	15	+ 3	29	7	+ 51
13–19	7	14	0	− 30	21	1	− 17	31	25	+ 45
20–26	8	10	8	− 32	20	10	− 27	24	20	+ 41
27–Mar. 5	9	13	2	− 43	20	11	− 36	24	39	+ 56
Mar. 6–12	10	14	8	−− 49	20	18	−− 38	29	42	+ 69
13–19	11	10	12	− 47	15	35	− 18	26	43	+ 86
20–26	12	12	1	− 58	17	14	− 21	24	40	+102
27–Ap. 2	13	9	10	− 57	19	14	− 26	21	30	+111
Ap. 3– 9	14	8	1	− 64	14	1	− 39	20	12	+103
10–16	15	9	13	− 60	14	9	− 44	22	24	+105
17–23	16	8	11	− 57	13	10	− 47	17	9	+ 97
24–30	17	14	2	− 69	19	1	− 65	17	1	+ 81
May 1–7	18	10	16	− 63	18	15	− 68	18	19	+ 82
8–14	19	9	6	− 66	13	17	− 64	17	8	+ 73
15–21	20	11	1	− 76	14	3	− 75	16	25	+ 82
22–28	21	11	3	− 84	13	8	− 80	13	16	+ 85
29–June 4	22	10	8	− 86	15	14	− 81	15	19	+ 89
June 5–11	23	14	0	− 100	14	1	− 94	12	11	+ 88
12–18	24	10	0	− 110	12	2	− 104	16	8	+ 80
19–25	25	11	1	− 120	21	0	− 125	19	15	+ 76
26–July 2	26	10	0	− 130	15	5	− 135	17	0	+ 59
July 3– 9	27	11	0	− 141	14	1	− 148	19	7	+ 47
10–16	28	11	3	− 149	18	19	− 147	17	2	+ 32
17–23	29	13	1	− 161	16	7	− 156	23	31	+ 40
24–30	30	15	5	− 171	23	71	− 108	17	37	+ 60
31–Aug. 6	31	10	9	− 172	20	57	− 71	22	31	+ 69
Aug. 7–13	32	10	8	− 174	22	28	− 65	22	20	+ 67
14–20	33	13	12	− 175	21	13	− 73	23	11	+ 55
21–27	34	15	5	− 185	25	6	− 92	23	18	+ 50
28–Sept. 3	35	17	4	− 198	20	17	− 95	32	24	+ 42
Sept. 4–10	36	15	2	− 211	17	13	− 99	23	29	+ 48
11–17	37	8	22	− 197	13	25	− 87	21	31	+ 58
18–24	38	10	1	− 206	17	2	− 102	17	16	+ 57
25–Oct. 1	39	15	0	− 221	19	10	−− 111	26	7	+ 38

It will be seen that all three districts began with an excess. Scotland North kept up an excess all through and finished September with 38 mm. (an inch and a half) to the good. Ireland South got on to the negative side first and kept it up with practically the same deficit as England S.E., about 160 mm., to the third week in July, then it made up more than half its loss while the English loss went on growing and at the end of September Ireland South was 111 mm. short, whereas England was down to the extent of 221 mm. The loss of 221 mm. is a serious matter; it is not far short of half of our expectation of rainfall, 561 mm. The experience has demonstrated that we cannot well get on with only half our normal rainfall and should make us consider what we are prepared for in the way of drought; a little further push in that direction this year would have been disastrous.

The line of maximum defect lies obviously east and west. Rains have come down on the northern side as far south as the middle of England, but have not come beyond.

The last memorable hot year was 1911, which evolved a temperature of 100° F. at Greenwich on August 11. It gave a very brilliant summer, but it was not so remarkable for dryness. At the end of September the defect of rainfall from the normal was only 141 mm.

The Peculiar Weather

Everyone is curious to know the reason for this exceptional behaviour of the weather in the south-east of England. It seldom rained, sometimes it threatened to rain and after almost promising to do so cleared up; and when it did actually rain, when past experience would have justified an expectation of violent thunderstorms and a prolonged spell of rain, it rained comparatively little and cleared up much sooner than we expected or even wished. It seemed as though the usual forces had lost control and the weather needed and indeed would be the better for the guiding hand of man. The offer to take over the control was made but not accepted.

Towards the Explanation

This is not the time to give a considered opinion as to whether the experience of this year will help towards the explanation of the facts of the distribution of rainfall over the Globe. Exceptional occasions often throw light on complicated problems such as this, but we have not yet the pertinent facts. The Meteorological Office gives us about 2500 separate facts daily about the weather of the British Isles. That means a good deal but we want to know about other places so that whoever will make a coherent explanation of the events will need a considerable capacity for assimilating and co-ordinating facts. The Meteorological Office seems to have grasped the significance of the situation for they anticipated the course of events ten days ahead. But they have made no communication as to any new form of procedure so we are left to make our own inferences as to whether their comprehension goes as far as explanation.

We may however survey the problem from a distance and put the problem in a more intelligible form than the mere statement of facts.

The Summer Circulation in the Northern Hemisphere

In order to form some sort of picture of what has been happening during
this droughty period it is best to have in mind the inference which can be
drawn from the maps of the normal distribution of pressure for the several
months. Take July as a typical example of the summer months of the northern
hemisphere. The general course of the air is in circulation round two areas
of high pressure and two areas of low pressure, which with their margins
cover the hemisphere (Fig. 43). Moving air always keeps high pressure on its
right hand, so it makes right-handed turns in circulating round a region of
high pressure and left-handed turns as it goes round regions of low pressure.

The Monsoon and its Return Currents

On the normal map for July (Fig. 43, plate XXIV) there is a conspicuous but
narrow region of low pressure under the Himalayas, extending from the
Persian Gulf to Assam, and round this goes a vast left-handed circulation
covering the whole of Asia, and its seas; the southern part of it, on its eastward journey, forms the south-west monsoon and a supply of eastward
moving air passes on as far as the Philippine Islands, turns northward and
then westward across Siberia to take its turn across the end of the Mediterranean and down the Red Sea to become the monsoon again. It is a vast
circulation. North of the Himalayas the whole continent is covered by a drift
from the north-east which floods the eastern Mediterranean with Asiatic
air. The wind in the extreme eastern margin of this great circulation passes
towards the pole along the Behring Strait between Asia and America and comes
down into the Atlantic as a flood of cold air from the west of Greenland.

The great South-Westerlies of the North Atlantic

Embracing Greenland itself is another area of low pressure, not so im-
pressive as the monsoon area but still of extreme significance for our
weather. On the eastern side of that apparently mild "low," between Iceland
and Norway is another vast current. It keeps the low pressure of Greenland
on its left and pours vast quantities of air from the south-west up to the
Arctic regions about Spitsbergen. That is the stream of air intermittent in
detail but perennial in effect which keeps the coasts of Norway, with Mur-
mansk, free of ice the year round.

The Intervening "High"

Between the south-westerly stream of the Greenland circulation over the
north Atlantic and the north-easterly stream over Asia of the return current
of the monsoon circulation there is a region of high pressure. It is very
fluctuating and irregular in its habits. It is too ill-defined to be easily represented; but it is always somewhere. The Baltic, or a little south of it, is one

of its favourite haunts, and from the Baltic it often puts out a tongue of high pressure towards an opposite tongue of high pressure which in the summer forms an extension of the most pronounced feature of the meteorological world of July, the great Atlantic anticyclone.

The great Anticyclones of the Northern Atlantic and Pacific

It is centred to the south-west of the Azores and has a well-defined marginal outline from the mouth of the Mississippi across the United States, along the steamer track to Bristol and thence to the mouth of the Thames, across by the Ostend route to Genoa and Tunis, and then back across Africa, the tropical Atlantic and the West Indian Islands. Thus the well-defined central region itself covers an enormous area. Another corresponding high pressure over the Pacific Ocean fills up the general scheme of circulation in the northern hemisphere, two lows, one over India, one over Greenland, with a belt of "high" between them which also forms a bridge connecting the two great "high" concentrations of the Atlantic and the Pacific.

Living on the Bridge

We have been living all this summer on or under that bridge. It is a sort of cantilever bridge with one base in the Atlantic and the other in the Pacific, and a very long and very variable connecting piece between the two. That connecting piece may be said to run roughly along the general coast-line of Europe and Asia, from the Belgian coast along the Baltic to the White Sea, and so on to the Pacific. It is subject to endless variations of infinite variety and it is perhaps overbold to sketch it as a recognised feature; but what has been happening all this summer is that the nearly permanent tongue of high pressure from the Azores has been very well marked; it has repeatedly been faced by another tongue of "high" coming from the Baltic, the two tongues have touched each other over and over again and formed a long belt of high pressure extending from beyond the Azores along the main coast-line of Europe to the confines of our daily maps; and quite frequently, over the British Isles as the point of contact of these two cantilever tongues, substantial erections of high pressure, local anticyclones, have been formed over and over again. They have been persistent far beyond use and wont.

Invasions from the North-West

The belt of "high" which is quite a natural consequence of the general circulation as represented on the map is in fact constantly open to attack by the stream of air which, tired of being the eastern fringe of the monsoon, has gone over the polar region and come down west of Greenland. One of the intriguing mysteries about air currents is the difference of their behaviour when they have to pass one another. If they use the British rule of the road and "keep to the left," they pass one another amicably, an anticyclonic area is set up, and the finest of fine weather reigns between; it is the result of that dispensation that we have been enjoying all the summer. But when the currents pass one another according to the continental rule of the road and "keep to the right," no sooner do they come within range, as they are accustomed to do to the north of us, than they curl themselves up into all sorts of violent contortions and the worst of weather is the result. That is what generally happens when the wind from the west of Greenland meets the south-west winds of the Atlantic. The two together form local "low pressures" or cyclones which pass along south of Iceland, often across our islands, not infrequently as far south as the Channel and crossing the North Sea go up the Norwegian coast-line. They have been doing that all this summer, with even more than their customary regularity, so that the north of Norway may complain of being too much according to rule and getting too much water. But this year fewer of the cyclones thus formed have crossed the south of England, they have seldom gone up the Channel, they showed an example of what they might do in the little gale of Sunday night, September 11; only on some occasions, as on August 16 to 18 or the following week, they crossed the Channel as it were by a supreme effort which resulted in one of the north-western depressions going to the south-east, thus crossing the "col" or saddle of high pressure instead of running along it as it ought to do. The orientation of the opposing "highs" and "lows" has changed, but even then it has generally left us in the "col."

The persistence of 1921

It may be asked why the conditions should be so persistent or recurrent in the year 1921? and the meteorologist's natural answer is, "why not?" It is what they are accustomed to further south, and a little north or south does not matter much. It is perhaps not unfair to say that if the science of the weather could be reduced to a few simple principles we should have the world divided into places where it is always raining and places where it never rains. It is variation rather than persistence that forms the crown of difficulty in the explanation—we would sooner face the persistence of this year than the variation of an ordinary year. It is quite possible that the persistence of weather in the current year may help us on the right way to an answer to the question of "how" and the subsequent question of "why."

*   *   *

The Unknown Origin of Anticyclones

Anticyclones have been constantly forming over our heads and we are driven to inquire how and why anticyclones form. Hitherto for the most part cyclones or cyclonic depressions have been our worry; we have tried to find an explanation of their origin and have left anticyclones to explain themselves. And yet anyone who aspires to control the weather may well say "if you will let me have the 'highs' you may do what you like with the 'lows'."

And an anticyclone is such a curious creature. What we know about it is that its store of air is always running away at the bottom as if it were in a leaky barrel, yet it remains and even grows. Clearly it must somehow or other be fed from the top, or at least some upper part; and when we try to think out the conditions of weather of this droughty year we shall find that the south-east of England was in a special sense the meeting place of winds which are represented by the noses of two anticyclones facing one another. As these two projections aligned themselves along the general European coast-line we got winds from the south-east and the north-west and had to supply air to the north-east and the south-west. We happen to have kept the bank of air all the summer; meteorologically speaking it is a risky thing to do because the conditions are not unfavourable for thunderstorms, only this year they did not come when they might have done. Perhaps the temperature or humidity of the air may have been exceptional. We may settle that point later but anyway the balance of air was in our favour, we got more air than we gave out and the anticyclone was the evidence. It must have been the upper air that did it because there certainly was leakage at the bottom, and if, in due course, we can get sufficient information about the upper air of this dry summer we may discover the secret of the formation of local anticyclones and secure an important step in the general explanation of our weather.

Let us briefly review our knowledge of anticyclones. On the maps of the distribution of pressure in the northern hemisphere (Figs. 28, 29, plates XX, XXI) we find first the permanent "highs" of the Atlantic and Pacific Oceans with their central regions about latitude 30° N. These are most pronounced in summer. Then there is the great seasonal "high" of winter over Asia, with a large central area of pressure normally above 1035 mb. extending over fifteen degrees of longitude east and west of 100° E., and about eight degrees of latitude on either side of 47° N. Its margin extends irregularly westward to the Pyrenees and north-eastward to the Behring Strait so that the line of 1020 mb. includes practically the whole of Asia and Europe. It is separated from an elongated belt of high pressure, extending from Egypt across the Atlantic south of the Azores to the Mississippi, by a band of low pressure along the Mediterranean. From the middle of the United States a high ridge passes to the Arctic Sea, separates the normal low pressure areas of the Atlantic and Pacific and forms a cantilever to connect with the northern side of the great Asiatic "high." Thus in the winter we find the most conspicuous region of high pressure is associated with the great cold of Central Asia. On its southern side is an irregular line of passage for cold air from the regions of the Caspian and Black Seas across Central Europe. In the fluctuations of the western margin of this area of high pressure our islands are sometimes brought within the range of the cold stream of air, and in that way we get our periods of winter frosts.

This great anticyclonic system is so clearly associated with the winter cold of Central and Northern Asia that we cannot resist the invitation to recognise in the cold, with its attendant high density of the air, the cause of the great anticyclone. It may be so if we consider the vast amount of cold air that must flow downward from the great Asiatic highlands in winter, and the persistent loss of heat from the ground by radiation. If we can regard the cold air at the surface as carried upward by the turbulent motion we may account for a great volume of cold air sufficiently thick to give the observed pressure at the surface. The explanation is at least plausible because in the great anticyclone the temperature increases with height, and the computations show that in the upper air at 4000 metres (Fig. 24, plate XIV) there is no longer any anticyclone, it is covered by the general cyclonic circulation round the polar axis. In like manner we may explain the anticyclonic conditions of Canada in winter, and a local anticyclone over Spain, by cold surface air stirred up to give a heavy layer. A part of the process may be that, in the absence of disturbing causes, air flowing downward from a cold region is deviated into an anticyclonic circulation by the rotation of the earth and the circulation tends to impound the air within it, leaving only a leakage at the ground.

In like manner the increase of the intensity of the oceanic anticyclones in the summer has been regarded as the expression of the relative coolness of the sea as compared with the neighbouring land, but there we are not quite on sure ground.

In any case the travelling anticyclones of our Islands are different. For there it has been ascertained that up to the level of 8000 metres the air in high pressure is relatively warm and in low pressure relatively cold. Above the level of 8000 metres things are again different, high pressure is cold and low pressure is warm. It appears that the anticyclones which control our weather derive their pressure from the coldness of the air in the highest levels, not from any coldness at the surface.

The observations of the upper air in this country have given at least one example of the process of formation of high pressure represented by the increase of pressure over England by two-tenths of an inch (6 mb.) between the 27th and 29th of July, 1908, as shown in Figs. 23 and 66. A consideration of the case may guide us in forming general ideas. The process seems to have consisted in the spread of a layer of cold air from the south which deranged the isothermal surfaces from being nearly vertical on the 27th to displaying a structure with a remarkable inversion on the 29th. The cold air seems to have spread itself between the warmer air above and below. Numerically from the nature of the calculation of height the explanation works quite well, though one cannot expect that it could do otherwise, but the change in the shape of the isothermal sheets is undeniable and most suggestive.

We are faced with the question, where on that occasion did the intrusive cold air come from? Presumably it came from below. In the stratosphere where the isothermal layers are vertical it could not come from above.

In this connection it is to be noted that the height which rising air will reach depends absolutely upon its conditions as regards temperature and moisture in relation to its environment; given suitable air and suitable environment the ultimate destination is inexorable, when it gets to the appropriate level it can neither go further nor go back: however much is delivered there must be accommodated. If we consider the process indicated in Fig. 44 we may suppose the position of the air when it reaches the level of equilibrium to be hundreds of kilometres from the locality where the convexion began. In suitable circumstances the air which is removed from one locality is disposed of, or dumped, over another at a great distance. The reduction of pressure in one part has for its counterpart the increase of pressure in another.

Further advance in this direction depends upon improving our knowledge of the course which the rising air follows. It is scarcely rash to say that whoever devises the means of tracing the trajectories of air in the three dimensions during the process of convexion will be able to give an adequate physical explanation of the formation of the anticyclones of middle latitudes.

7. THE CYCLONIC DEPRESSIONS OF MIDDLE LATITUDES

I. NOTE ON A CONFERENCE OF NORWEGIAN AND BRITISH METEOROLOGISTS AT BERGEN, July 19–30, 1920

COMMUNICATED TO THE SUB-COMMITTEE FOR METEOROLOGY, NAVIGATION AND ATMOSPHERIC ELECTRICITY OF THE AERONAUTICAL RESEARCH COMMITTEE

1. *A new method of forecasting by the detailed study of the polar front in cyclonic areas.* With the sanction of the Air Ministry the Meteorological Office accepted an invitation from Professor V. Bjerknes, Director of the B Division of the Geophysical Institute of Bergen, to visit Bergen for a conference with Scandinavian meteorologists.

The purpose of the Conference was to hear and discuss accounts of the new analysis of the phenomena of cyclones of middle northern latitudes which had been worked out at the Bergen Institute under Professor V. Bjerknes' direction, and had been introduced into practical meteorology as a new method of forecasting principally by Solberg, J. Bjerknes and Bergeron.

The method originated as the result of the "microscopic" study of the distribution of pressure, temperature and rain over Norway which was adopted during the war as a substitute for the observations from the wider area for which the daily maps were compiled in peace time. Its first conclusion was the differentiation of the air over Norway into two distinct parts separated by a line of discontinuity of wind-direction, temperature and humidity. The air on one side of the discontinuity, on the northern or western side, is cold, dry and transparent, while the immediately adjacent air on the southern or eastern side is relatively warmer, moister and less transparent. The fact of its being colder and drier on the northern side suggests a polar origin for the air and the process is looked upon as part of the alternate advance and retreat of a "polar front" of which the line of discontinuity at the earth's surface marks the edge.

2. *Reference of the phenomena to the two datum lines, "steering line" and "squall line."* The line of discontinuity or edge of the polar front has a definite relation to cyclones with marked centres. From the centre of each cyclone along a line, which is called the "steering line" because it starts from the centre in the direction in which the cyclone is moving though it may curve to the southward further from the centre, the polar front is regarded as being a bank of air with stream lines to the south-west, west or north-west over which the equatorial air is advancing gradually upward by motion directly transverse to the line of motion in the front. We have called this the anaphalanx of the polar front. It is now the *warm front* of J. Bjerknes.

Also from the centre of the cyclone along a line initially at right angles to the line of motion of the centre the air of the polar front, having curled round the centre from the east or come into the area as a north-wester, undercuts the flank of the equatorial air advancing from the southward towards the anaphalanx and throws it upward. The surface boundary of the polar front in this region is called the "squall line" and the advancing surface the kataphalanx of the polar front[1]. With these two sections of the front rainfall is associated in a manner that can be easily generalised and it is the association with rainfall that makes the suggestion specially effective for forecasting. Many of the phenomena of the squall line or the kataphalanx have been discussed in the papers by Lempfert and Corless on line squalls.

3. *The warm sector.* Of the complete area of the surface section of a cyclone a sector of 90° at the centre but of varying angle further out and always less than half the area of the cyclone, is occupied by "equatorial" or warm air destined to move up the anaphalanx or to be undercut by the kataphalanx.

For want of a better name this sector is called the "warm sector." Thus the area of each cyclonic depression is mapped without much difficulty by lines meeting in the centre and dividing the warm sector of about 90° from the remaining 270° of the polar front. The weather in the warm sector may be showery but it is not literally rainy. The properties of each cyclone are referred in practice to these two lines instead of to the centre which is the traditional point of reference for the properties of a cyclone.

4. *The section of the polar front in the region between two cyclones.* When there is a succession of cyclones the position of the polar front can be marked in the interior region of each of them and it must necessarily have some boundary in the intermediate region between the successive cyclones. This part of the boundary is not so easily identified as the other parts. Such intermediate regions are represented on our maps generally by "wedges" or protuberances northward of high pressure from the margin of the permanent tropical anticyclone or its extension eastward. So the boundary of the polar front between cyclones finds itself or loses itself somewhere in the prevalent westerly current of latitudes 40° to 50°. This may be represented by a region of "straight isobars" which belong partly to the polar front and partly to the equatorial air which has come round the high pressure from the south-west. Thus the region of prevailing westerlies may be a region of oscillation of the polar front between the ridge line of the tropical high and the outer margin of that high.

But it is obvious that any polar air which forms the eastern margin of the Atlantic high pressure may, and occasionally must, extend southward to the equatorial region with the current which penetrates the line of highs as the north-east trade-wind. Hence the polar front may, and occasionally must, find itself ultimately to be feeding from the north-east a westward moving equatorial current. With possible interruptions of that kind, the polar front may be traced round the world by a boundary which marks off a cold, dry, transparent polar cap from the rest of the air of the northern hemisphere. A similar boundary will be found in the southern hemisphere.

5. *The polar front as representing Helmholtz's surface of discontinuity and Bjerknes's suggestion of wave-motion as representing the cyclonic depressions of middle latitudes.* Professor Bjerknes suggests the line of discontinuity between

---

[1] J. Bjerknes's more recent name is the *cold front*.

the polar cap and the equatorial air as the locus of cyclonic changes and, following a suggestion of Helmholtz that on a rotating sphere such a surface of discontinuity would tend not to be vertical (except at the poles) but to be parallel to the polar axis, he regards the polar front as the expression of Helmholtz's discontinuity and proposes something analogous to wave-motion in the surface of discontinuity, different on the two sides, as a dynamical effect of which the horizontal projection appears on our maps as more or less centrical cyclonic depressions.

So he would regard the surface of discontinuity as a general dynamical effect of the earth's rotation upon air of different velocities, densities or gravities, and therefore unlimited as regards height in the atmosphere irrespective of the source from which the coldness, or other cause of change of specific weight which is the key to the discontinuity, is produced. Helmholtz's theory was discussed and a good deal amplified or magnified by Brillouin in *Vents contigus et Nuages*[1], and applied by him to the explanation of the forms of clouds. Helmholtz's deduction is however of the same order as the deduction that moist air will become stratified above dry air because it is specifically lighter, or that the percentage composition of the atmosphere will change with height. It postulates undisturbed conditions in which the air can arrange itself according to the principle to be applied.

I am therefore doubtful whether it is possible to treat actual details by a rigorous dynamical method on account of the assumptions that have to be made in order to reduce the questions to algebraical form. For example, wave-motion is periodic in a very special manner. One begins by supposing a train of waves. The phenomena of cyclones are also periodic, but in a rather different sense and they need a good deal of analysis before the periodicity appropriate to wave-motion emerges. Professor Bjerknes is, however, more conversant with the peculiar dynamical properties of fluids than anyone else and he may see his way through the difficulties and produce on the hypothesis selected an explanation of the realities of depressions[2].

6. *An alternative view of the polar front as representing the boundary of a cap of polar air fed by convexion of cold air from high land in polar regions.* Meanwhile the subject may be viewed from a slightly different point of view with some interesting features behind which also there must be general dynamico-physical principles that are not yet explored. It depends upon an analysis of the general circulation of the atmosphere into two parts by a horizontal cut at 4000 metres. The lower part is a cap of cold air which may be attributed to the loss of heat by radiation from the surface of land in the polar region.

7. *Distribution of pressure due to distribution of temperature in the surface-layer and the east to west circulation of air which would correspond therewith.* I showed in 1904[3] from Teisserenc de Bort's isobars for 4000 metres (Figs.

[1] *Ann. bur. cent. météor.* 1896, part I, pp. B 45–150, Paris, 1898.

[2] Professor Bjerknes's analysis is now published (1922) under the title "On the dynamics of the circular vortex with applications to the atmosphere and atmospheric vortex and wave-motions," *Geofysiske Pub.* vol. II, no. 4, pp. 1–88, Kristiania, 1921.

[3] *Proc. Roy. Soc.* vol. LXXIV, p. 20.

24–27, plates XIV–XVII), that the atmosphere could be divided into two parts with opposite rotational properties by a cut at 4000 metres: the part below 4000 metres shows a distribution of pressure strictly corresponding with, and derived from, the surface distribution of temperature. That temperature is high in the equatorial zone and low in the polar regions, and the corresponding pressure would be high in the polar regions and low at the equator and therefore would be represented by a general circulation of air round the poles from east to west. We find such a circulation of air in the equatorial region between the tropics and may surmise that the easterly equatorial current is itself an extension of the polar front. In the northern hemisphere the distribution is less regular than in the southern because of the disturbing influence of the irregularities of land and water.

For this distribution, temperature is directly responsible and the phenomena are most conspicuously developed in the winter of the northern hemisphere when the loss of heat by radiation from high land is greatest, particularly in Greenland and the higher regions of America, Europe and Asia. The centre of the circulation is, however, in Eastern Siberia, not at the North pole. From the physical point of view the polar front is the temporary boundary of this mass of cold air which is endowed with circulation from east to west and is gradually spreading itself towards the equator from both sides.

8. *Extension to explain the origin of tropical anticyclones.* If the approach towards the equator could be distributed equally round the whole equatorial zone the polar components of the opposing streams should make high pressure there: but, instead of that, parts of the front penetrate towards the equator faster and more easily than the rest, and they converge from each side to form an equatorial current from the east which imprisons masses of air on either side and thus forms isolated patches of tropical high pressure with an intervening belt of low pressure instead of the single equatorial high.

9. *The supposed descent of air in anticyclones.* This is probably the correct explanation of the permanent anticyclones of tropical oceans. The commonly accepted explanation of the maintenance of these anticyclones by the descent of air cooled by radiation as it flows northward from the equator, after being in the doldrums, is untenable for two reasons. First, it is potentially the warmest air of the world and is subject to intense solar radiation: it is therefore the least likely of any air in the world to be cooled enough to descend spontaneously. Secondly, the descent of air in the anticyclones is at about the rate of a millimetre per second or 86 metres in a day[1]. Thus these great

---

[1] We can now make an approximate calculation of the descent of air over a region like the high pressure of the North Atlantic Ocean in summer, regarding it modestly as a circular area with a radius r of about 10° of arc, or 1000 kilometres. We take the angle of deviation of flow from the isobar to be a, V_H to be the horizontal surface velocity, and V_Z to be the downward velocity of convexion. We may take convexion to be uniform over the area, because the surface-wind increases with the distance from the centre, perhaps not in strict proportion but not very differently therefrom.

Further we require an estimate of the height to which the outflow extends, and for that purpose we shall suppose that by uniform stages agreement with the gradient is reached at a height of 500 m., and that the mean velocity of outflow over the vertical distance of 500 m.

anticyclones are truly regions of descending air if the month or the year be taken as the unit of time, but for the units of time which are appropriate to dynamical questions, namely the hour and the day, they are simply quiescent regions of great stability where the ordinary physical operations can go on comparatively undisturbed by changes in the environment. The same is generally true of all anticyclones; dynamically they are very stable and undisturbed although a small amount of air is withdrawn from the bottom layer across the marginal isobars which might bring the top air to the bottom in about one-third of a year.

It is this transfer of air along the surface across the isobars and the resulting subsidence of air in anticyclonic regions which represents a large part of the compensation for the air which is being lost to the surface by convexion. As a general principle we may conclude that the convexion of ascending air is localised and rapid, being completed in hours or perhaps in minutes, whereas the descent is by gradual subsidence over anticyclonic areas which would require months for completion, except in the regions of high land, where, under the direct influence of terrestrial radiation or reserve of cold represented by permanent ice or snow, there are local cataracts of air which may be described as katabatic winds.

10. *Reversal of the distribution of pressure and consequent circulation between two and four kilometres. The limitation of the two circulations and their combination to produce the actual distribution of pressure and wind at the surface.* The layer above the cut at 4000 metres exhibits the curious effect of a circulation from west to east round the poles in which high pressure and high temperature go together. The great rotating cap which is represented by this distribution of pressure extends from about 30° of latitude to the poles. It does not (as a rule) challenge at its own levels the equatorial belt of easterly winds fed by the polar front. That belt which, with the notable exception of the south-west monsoon in the summer, occupies the surface from about 30° N. to 30° S. extends to all heights and probably represents about one-half of the air of the globe. But it is not entirely homogeneous; it is certainly invaded by monsoons and also by other special circumstances.

The characteristic features of the westerly circulation are fully borne out

is one-half of the outflow at the bottom. That being $V_{II} \sin a$ the average velocity of outflow may be taken as $\frac{1}{2}V_H \sin a$.

Whence we obtain for the outward flow $\pi r h \sin a V_{II}$ and for the downward convexion necessary to supply this flow $\pi r^2 V_Z$. These are equal, hence

$$\pi r^2 V_Z = \pi r h \sin a V_H \quad \text{or} \quad V_Z = \frac{h \sin a}{r} V_{II}.$$

If we insert the numerical values (taking 1 metre as the unit of length) we get:

$$V_Z = \frac{500}{1{,}000{,}000} \sin a . V_{II} = 5 \times 10^{-4} \sin a . V_{II}.$$

If the angle of deviation a be 30°, and the horizontal wind 4 m/s (about 10 miles per hour) the vertical velocity works out at 10^{-3} m/s, or one millimetre per second. This means a descent over the whole area at the rate of 3·6 metres in an hour, or 86 metres in a day. At that rate it would take a hundred days to bring the air down from 8600 metres to the surface. (From the introduction to a memoir on Tropical Hurricanes, M.O. Publication, 220 j.)

by observations of the motion of upper clouds; and the most remarkable feature of the circulation is that the lines of pressure-distribution which mark it are extraordinarily similar, even in such details as are exhibited on a map of small scale, to the lines of pressure-distribution for the lower four thousand metres which are derived expressly from the surface-temperature.

We find therefore that the gradients of the distribution of pressure in the upper air (above 4000 metres) are almost exactly the inverse of those which correspond with the distribution of temperature at the surface. The actual distribution of pressure at the surface is the resultant of the superposition of the two opposites and picks out the differences of gradient in the two as isolated anticyclones and regions of low average pressure.

The inversion takes place somewhere between two kilometres and four kilometres. About one-third of the atmosphere belongs to the lower *régime*.

11. *The westerly circulation of the upper air attributed to the reversal between 7 and 17 kilometres of the distribution of temperature prevailing at the surface so that in the uppermost layer the equatorial region displays the lowest observed temperatures.* And this inversion of pressure-distribution from opposing gradients of pressure and temperature to concurrent gradients may in a manner be regarded as the sequel of another and more direct inversion of temperature-distribution which takes place between 7 and 17 kilometres with about one-tenth of the atmosphere remaining above it, and again concerns itself with about one-third of the atmosphere. As a result of this second inversion the distribution of temperature is the reverse of that at the surface. At those levels air is coldest in the equatorial zone and warmest in the polar regions. The corresponding pressure would be high at the equator and low at the poles. As in the case of the partial pressure of the surface layer the distribution of pressure may again be regarded as the statical result of the distribution of temperature.

It is possible that other inversions may or must take place still higher up. It is not proposed to speculate further upon the formation of the two inversions that are already established, namely, the inversion of pressure-distribution between two and four kilometres and the inversion of temperature-distribution between 7 and 17 kilometres or upon the general application of such inversions. What is *à propos*, however, here is that from this point of view the discontinuity between the polar front and the equatorial air is an incident in the flow of cold air from cold highlands towards the equator. It is constantly changing its position and shape and probably is nowhere left undisturbed long enough for the Helmholtz surface of discontinuity to be developed through any great height from the surface.

12. *Circulation due to local convexion as an alternative explanation of secondary cyclonic depressions in lieu of the wave-motion which is not yet fully analysed.* What then is the alternative to wave-motion in the polar front as the origin of the local irregularities of the distribution of pressure in middle latitudes? We propose to consider the efficacy of local convexion as a cause of cyclonic circulation in the upper air.

13. *The late Lord Rayleigh's conclusion that cyclonic motion is due to the "annihilation" or "eviction" of fluid in the interior of a mass of air which already possesses some "vorticity" or relative motion.* In the late Lord Rayleigh's analysis of revolving fluid[1] it is the "annihilation" of fluid within a region of existing vorticity or relative motion that causes the development of a simple vortex. In order to avoid any misunderstanding from the use of a word which would apparently defy the equation of continuity the reduction of mass as the incidental result of convexion will be called the "eviction" of air.

The advantage of analysis for meteorology is that it enables us to bring the actual facts derived from observation into comparison with the conditions which can be computed on the one hypothesis or the other, which is a preliminary in the identification of the forces and conditions upon which the actual situation depends.

14. *Objections to the commonly accepted view of the rôle of convexion and suggestion of a new principle of action of convexion, namely, the eviction of air by the scouring action of an air-current in the layers through which the convexion passes instead of the simple replacement of air in those layers by potentially warmer or colder air.* Years ago the effects were attributed directly to the convexion of warm air giving rise through its small density to cyclonic areas of diminished pressure with roughly circular isobars. That theory has been entirely discredited in recent years because the reduction of pressure was attributed to the replacement of the air of the central region by warmer, and therefore lighter, air from the surface, whereas it is agreed on all hands that at any considerable height above the surface low pressure is cold and high pressure is warm.

In working recently at the phenomena of tropical revolving storms I have come to regard the process of convexion from a new point of view which may have a number of applications in meteorological theory. The practice of regarding the result of thermal convexion as the permanent replacement of air by warmer air from below seems to be unnecessary as well as inappropriate. As regards any particular layer the presence of a mass of warm air beneath it, regarded as a balloon with sufficient buoyancy to go through it, would mean the pushing aside of the air above as the balloon came through and the closing in again after it had passed. After oscillations had subsided the whole process would leave the layer just as it was before except that the surface friction of the balloon would have enabled it to carry away with it some air which properly belonged to the layer.

If a similar process be repeated with innumerable balloons, or still better with innumerable masses of air, the effect will be that the rising air will drag up with it from the layers through which it passes a considerable amount of air by eddy-mixing between itself and the environment: such mixing may retard but it cannot suppress the ascending motion due to difference of temperature between the air and its environment because the temperature of the

---

[1] *Proc. Roy. Soc.* A, vol. XCIII, p. 148, 1917.

environment determines the buoyancy; and the mixing of the buoyant mass with its environment cannot reduce its buoyancy to zero.

Hence we may regard convexion as the dragging upwards of a portion of the layers *through* which the air passes on its way up. Convexion may in this way remove from any layer a mass of air more than equal to the mass of air which has travelled through the layer, the removal of the air being due not to its intrinsic buoyancy but to the scouring action of other buoyant air which passes upward till it finds itself in equilibrium with its environment at some higher level.

The operation of convexion of air from this point of view may be expressed as the eviction of a quantity of air belonging to the layers traversed by the convexion: according to the late Lord Rayleigh's showing, the eviction of air in a region where there is already rotational motion is followed by the superposition of a simple vortex upon the original motion of the environment.

Thus we may have convexion accountable for the development of revolving fluid and it must be noticed that the rotation of the fluid is developed by the convergence, in a region with some existing vorticity, to the locality from which air has been evicted. The convergence will be attended by reduction of pressure and consequent automatic reduction of temperature: so that the interior region of the consequent cyclone will certainly be cold. The locality of reduced pressure will be protected against further invasion by the rotation developed in the fluid, and when the convexion is all over we shall have a central region of low pressure and low temperature protected from collapse by its rotation and vulnerable only through the creeping in of fluid and other things at the bottom where friction prevents the maintenance of the velocity necessary for protection.

From this point of view I have traced the effect of convexion in tropical revolving storms in an official memoir, but let us now suppose that thermal convexion is by some means or other set up within the general westerly circulation at the 4000-metres level of the northern hemisphere, the inexorable result is eviction of fluid and a superposed vortex, in certain layers, which may be permanent or transient according to circumstances. Permanent if the vortex can be gradually extended upwards with diminished rotation and extended downwards till the ground is reached, and transient only if these conditions are not satisfied.

15. *The corresponding effect of falling rain or sand. Calculation of the convective effect of a cloud burst (Glossary, p. 334, No. 6) as a wind of ·72 m/s.* This view of the eviction of air as the indirect dynamical effect of the convexion of air passing through a layer can of course be extended to descending air, such as may be found in the kataphalanx of a cyclonic area, where, however, the eviction takes place in an extended line, and moreover it can be extended to falling particles of solid or liquid and therefore to falling raindrops. In most instances, no doubt, the dynamical effect of falling rain as an evictor of air in the layers through which it falls is insignificant, but heavy rain may be effectively operative.

The following is an estimate based on the tables in the article "Raindrops" (*Glossary*, pp. 334–6). The 200 drops per m² per sec. of No. 6, table 1, have a volume of 0·0478 c.c. or mass 0·0478 g. They are falling with their end velocity of 8 m/s. In one second they will have traversed 8 m. and the momentum 200 × 0·0478 × 981, which gravity might have given to them will have been communicated instead to 8 m³ of air of approximate density 1250 g. per m³, i.e. to 10,000 g. of air. That amount of momentum per second can be expressed as a certain acceleration a of the 8 m³ of air where (in C.G.S. units)

$$10,000 \, a = 200 \times 0\cdot0478 \times 981,$$

hence

$$a = \cdot94 \text{ cm/sec}^2.$$

The larger number of smaller raindrops which constitute the rainfall of column 6, referred to, may add 25 per cent. to this amount.

Hence the effect of the heavy rain upon the column of air through which it passed would be to superpose on the air, through which it was travelling with the end velocity of its drops, a downward acceleration of about 1·2 cm/s. If rainfall of that description continued for a minute and no allowance is made for the time required to maintain the end velocity in the downward-moving air, the result would be a velocity of ·72 m/s. The effect is enhanced by the reduction of temperature of the falling rain and probably the result of that is a destructive squall often the preliminary of a heavy downpour. Hence heavy rainfall may be effective in helping the eviction of fluid from the layers through which it passes. (See Appendix, p. 228.)

16. *Aitken's experimental development of a secondary in a column of revolving fluid by falling sand.* We may note here an experiment of Aitken's[1] wherein a "secondary" was observed to be formed in a body of rotating liquid by dropping sand into the liquid, following with the hand the point of fall to get a vertical column of water affected by the falling sand. The effect of the sand would appear to be the eviction of fluid in its path or, speaking dynamically, the production of a downward acceleration throughout a particular vertical column of the rotating fluid thereby causing diminished pressure first at the top and subsequently down the whole column. A local rotation of the column or secondary system of rotation in the fluid is the inevitable consequence.

17. *The co-operation of the combined effect of rainfall and convexion in evicting fluid and producing secondaries in a cyclonic depression.* It may now be remarked that strong vertical convexion of air itself results in heavy rain and consequently two causes of eviction of fluid act conjointly upon the column in which air is rising and rain is falling. Hence the creation of a partial vacuum is inevitable and the conditions necessary for the production of a superposed vortex of revolving fluid as a secondary in the existing circulation are sufficiently provided for.

With very strong convexion and very heavy rainfall the superposed vortex may be of great intensity; and here it may further be remarked that if the

---

[1] "Revolving Fluid in the Atmosphere," *Proc. Roy. Soc.* A, vol. XCIV, p. 250, 1918.

convexion and associated rainfall can be continued for a considerable time over a limited area which travels with the current the vortex will be of extreme intensity because the law of velocity is $vr = $ constant, the curve which represents it is a rectangular hyperbola. For a given eviction of fluid if the radius r is very small v will be correspondingly large.

18. *The primary depressions of the tropics. Computation of the amount of air evicted in the tropical revolving storm of Cocos Island.* What is noted here for secondary depressions applies equally to primaries formed in any current such as an equatorial current in which a certain amount of vorticity or local difference of velocity already exists; and it is a sufficient explanation of the origin of revolving storms in tropical regions. We can indeed calculate approximately the amount of air which must be removed in the development of a cyclone of given dimensions and intensity.

I have made a calculation of this kind for the cyclone of the Cocos Islands represented by a barogram on p. 156 of the *Meteorological Glossary* by taking the time-scale as representing distance at 16 kilometres to the hour and the graph of the barogram as indicating a section of a column of mercury standing upon the area covered by the storm. The result of the eviction of air at the epoch represented is shown by the loss of pressure over the area and is obtained by a step-by-step integration of the weight of mercury missing from below the datum line of the undisturbed region on either side of the centre. I have taken the line of 1015 mb. as datum. My calculation shows the volume of mercury to be 526/6292 of the original quantity of mercury above the line of 950 mb. and hence sufficient air must have been evicted to lower the pressure over the whole area by 526/6292 × 65 mb. or 5·5 mb. That corresponds with the removal of air equivalent to 5·5/120 of a kilometre, about 46 metres of air at sea-level, from the whole of the area which covers about 800,000 square kilometres or 460 metres from 80,000 square kilometres, about 40,000 cubic kilometres[1].

How much surface air would be required to cause the eviction of that amount it is not possible to say until we can estimate what the frictional effect of an upward current would be.

19. *The maintenance of existing systems and the deepening of the depression by convexion.* It may however be noticed that a great part of the effect is expended in reducing the pressure slightly over the whole of the area affected. When the area of convexion has become restricted to a comparatively narrow column very little additional removal may cause a notable further diminution of pressure.

This calculation is certainly not directly appropriate to all the cyclonic depressions of middle latitudes, but it may be pertinent to the life-history of those which begin as tropical revolving storms and travel round the tropical high pressure into the region of the prevailing westerlies.

The rotating system once established needs only little convexion for its

---

[1] The air might contain 625 million tons of water, equivalent to rainfall of 0·72 mm. over the whole area.

maintenance, which may be obtained so long as the surface air which creeps into the interior at the bottom is sufficiently warm and moist to be pushed upward by its environment within the ring of maximum velocity. When that is no longer the case the system must gradually lose intensity; this will certainly happen if the vortex travels over a layer of cold sea, and, for that or another reason, has to take in cold air. If it can find cold and warm air in juxtaposition it may profit by the convexion so that the edge of the polar front might be a favourable locality for the further persistence of an old cyclone.

20. *The possibility of experimental illustrations of the eviction of fluid by the convective action of heated air or gas upon air, or water drops upon air, or solid particles upon air.* The convexional process, which is here described as an alternative to wave-motion in the polar front by way of explanation of some of the phenomena of the cyclonic depressions of middle northern latitudes, could evidently be made with advantage the subject of experimental study.

The convexional effect of falling drops could be definitely measured, and the relation of size and weight of particles to the intensity of the secondary produced in a repetition of Aitken's experiment could be ascertained and the experiment could be pushed to the limit of absorption of the primary into the secondary.

The amount of air lifted from the environment by air in convexion could be determined by noting the area and temperature at different levels of a column of rising air produced by heating a limited surface, and a model vortical whirl might also be produced by convexion if a substitute for atmospheric stratification by pressure were introduced above the heated surface. Such stratification could probably be provided by a series of per-forated plates of sheet metal with the perforation not less in diameter than the natural width of a column of ascending air, and if that could be managed a number of experiments on the production of secondaries and other phenomena could be arranged within the limits of an ordinary room in a laboratory. Very few experiments of the production of a tornado by thermal convexion have been tried and modern facilities for localising heat by electric currents which are available would make things easy and definite. By en-larging the scale to some extent the laboratory experiments upon which the theories of meteorological phenomena have been supposed to depend could be revised by being brought into relation with modern knowledge of atmo-spheric structure. Investigations of this kind might very well accompany the further development of dynamical theory from the mathematical side.

21. *The eviction of fluid in Mr Dines's model of a tornado.* It may be noticed that in Mr Dines's apparatus, now in the Science Museum, for illustrating the formation of a tornado the air is evicted at the top by a fan-wheel in the ceiling and the original vorticity is imposed upon the air which approaches the centre to replace that evicted by suitably placed planes regulating the admission from the outside. The approach to the axis begins at the top and extends downwards as the top protects itself by rotation and the eviction continues in the interior. The development downward is very rapid.

22. *The structure of cyclonic depressions of middle northern latitudes.* The revival of the idea of revolving fluid developed by convexion as the dynamical basis of the structure of those cyclonic depressions which are carried from tropical to temperate regions has to face the general objection to supposing that any large masses of atmospheric air can be carried about the world with only slow changes of the material of which they are composed, and the specific objection which arises from the want of symmetry of the surface air of actual cyclonic depressions, as demonstrated in the *Life-History of Surface Air Currents*[1], and more recently expressed by the division of the surface into the warm sector and the polar front.

Having regard to this division, from which no exceptions are apparently allowed, it is certainly fair to say that the air of our cyclonic depressions is supplied from two perfectly distinct sources, one warm and moist and the other cold and dry, and therefore they offer no encouragement to a theory which postulates the persistent rotation of air round a central column.

Against these objections it must be said first, that in the case of September 10th, 1903, analysed in the *Manual of Meteorology*, Part IV, chap. XI, the corrected section is very suggestive of a hyperbolic vortex, and that the remarkably high coefficients of correlation between changes of pressure and temperature obtained by Mr Dines from observations in England, grouped only according to quarters of the year, is *primâ facie* evidence that, whatever may take place at the surface, from three kilometres to eight kilometres the symmetry appropriate to the persistent revolution of the same fluid is certainly disclosed.

Dr E. H. Chapman[2] has kindly examined for me the question of making a correction of the crude coefficients of correlation on account of the probable errors of observation which are quite independent of the intrinsic deviation from proportionality of corresponding changes. By an ingenious statistical operation he has obtained the statistical correction and given me the following.

TABLE OF CORRELATION COEFFICIENTS BETWEEN CHANGES OF PRESSURE AND CHANGES OF TEMPERATURE AT DIFFERENT LEVELS IN THE ATMOSPHERE CORRECTED FOR PROBABLE ERRORS OF OBSERVATION

Height in km.	2	3	4	5	6	7	8	9	10
Jan.–March	0·935	0·885	0·946	0·931	0·916	0·953	1·000	0·903	0·400
April–June	0·559	0·900	1·000	0·992	1·000	0·957	0·891	0·491	0·216
July–Sept.	0·638	0·817	0·848	0·903	0·913	0·957	0·957	0·959	0·464
Oct.–Dec.	0·866	0·862	0·913	0·953	0·935	0·931	0·986	0·870	0·328

23. *Change of structure below three kilometres.* These figures present a structure for the layers between three kilometres and eight kilometres, which is quite as obstinately in favour of symmetry as the conditions at the surface are against it. (See p. 101.)

---

[1] M.O. Publication 174, 1906.

[2] "The Relationship between Pressure and Temperature at the same level in the Free Atmosphere," *Proc. Roy. Soc.* A, vol. XCVIII, 1920, pp. 235–248.

And there is a further general tendency in favour of a stereotyped symmetrical form which arises from the fact that all dynamical operations upon air tend to result in reduction to a normal structure with the layers arranged in order of realised entropy. In this view water-vapour counts as potential entropy and the removal of the water-vapour by precipitation after condensation as the realisation of the entropy. The process of realisation in that sense is irreversible. Hence, if any layer becomes subject to diminution of pressure as, for example, by convergence in a cyclone, any pocket of exceptional local humidity, and therefore of potential entropy, will develop condensation and consequent instability, stability only being secured when the distribution has become regular.

There are, therefore, certain forces tending to regularity and symmetry of structure and it would appear from the correlation coefficients that the travelling changes of pressure yield to these forces in the layers between three and eight kilometres, but resist them at the surface so successfully that there is no relation between the variations of the two quantities in the lowest levels.

24. *The vicissitudes of travelling revolving fluid.* If we endeavour to form a picture of the life-history of a column of revolving fluid it appears indeed to be remarkable that it can hold together at all; and the fact that it does at least preserve the characteristic features in regard to pressure and winds at the surface, accompanied normally by symmetry of pressure and temperature at middle heights, and not infrequently by apparent circulation in like sense at great heights, indicates some stability of structure of which vortex-ring motion and wave-motion are the only types which experimental physics has as yet disclosed[1]. We can picture to ourselves many obvious reasons for the dissipation of the kinetic energy of a column of revolving fluid, but hardly any for its conservation.

In the equatorial region, for example, the differences of velocity of the easterly equatorial current at different heights suggest that convexion would fail to form a vortex on account of the shearing of the upper part from the lower by the relative motion of the environment. This must actually be the case so frequently that it would be interesting to examine the records of stations within the regions of formation of those meteors for evidence of abortive attempts at the formation of cyclonic whirls. The fact that such whirls are formed may be taken *primâ facie* as evidence that though the conditions favourable for development are certainly very rare, since fully-developed cyclones are not so common as the surface conditions which might cause them, yet occasionally they do occur.

So long as the vortex travels over the sea it has only the differences of velocity in the upper air and the leakage inwards of the lowest layers extending perhaps through one kilometre to contend with; but when it passes over land

[1] Vortex-ring motion has a variety of forms. I recall a remarkable experiment by Mrs Ayrton at a Meeting of the Royal Society to demonstrate the use of flappers. By "flapping" the lecturer's table two or three times a parcel of air was made to travel down the room with a perfectly definite velocity of a few feet per second. Presumably some sort of self-contained vortex was set up by flapping the table but it was not a ring in any ordinary sense.

of varying height its bottom is certainly cut off up to the height of any range of hills which it crosses and its interior is then exposed to the reception of a supply of new air, the effect of which may be gradually to degrade the kinetic energy of the whole column. It may recover itself with or without a reduction of its whole energy if the new supply of air is sufficiently warm and moist[1].

Having traversed a stretch of the equatorial region the column not infrequently travels round the western margin of an area of high pressure and transfers itself to the general westerly current of middle latitudes. Whether it does this in consequence of a movement of air round the high pressure approximately uniform at all heights or by depending upon its dynamical stability for the crossing of the zone intermediate between the easterly current and the westerly current is not known. And the phenomenon of its successful negotiation of this passage is the more remarkable in the case particularly of the West Indian hurricanes, because the equatorial current over the West Indies is subject to well-recognised irregularities. The ashes of eruption of the West Indian volcanoes are said to fall to the east of the points of ejection, not to the west. But these points lie on the south-west of the usual paths of hurricanes and the atmospheric structure may be complicated by the proximity of the continental area of South America.

And if, from conditions hitherto insufficiently explored, the travelling column is successfully launched on the general stream of the prevailing westerlies we have to reckon with the fact that the currents are often very irregular below the level of four kilometres, and the revolving column can only rely upon the strata above four kilometres even for the uniformity represented by a general average. We may fairly suppose that the perturbations of the stream above four kilometres are represented by a column of revolving air in those layers because of the symmetry which is characteristic of those layers; but for the lower regions the reasoning fails. The most plausible suggestion is that revolving columns preserve their structure in those upper layers while the motion of the layers beneath is controlled by a complexity of forces of which pressure-distribution transmitted downwards from above is only one element. In favour of this view we have the facts that from the observations of pilot balloons in the stratosphere of Europe according to G. M. B. Dobson there are as many north-easterly winds as south-westerly.

A revolving column in the general stream is also subject to two other perturbations. First by the nature of a vortex it requires for unimpaired structure the extension of its rotational effect over an indefinitely great area, the margins of which are not under its control, and it must be subject to gradual degradation on that account. Secondly, being represented by a rotating column of great area and small height it is exposed to the formation of local secondaries within its own body in the manner already explained, which may result in the transference of the main centre of rotation to some other point of its area and a fresh arrangement of material.

---

[1] The process of the passage of a cyclone over the mountains of the Corean peninsula is described in a paper by T. Kobayasi before the Royal Meteorological Society in February, 1922.

If it preserves its identity through all these adverse circumstances it may pass along the west coast of Norway as the survivor in the upper air of the column that started from the equatorial Atlantic and finally lose its energy entirely over the polar ice or conceivably, though not probably, circumnavigate the pole. All this time the only part of it that can in any way claim to be the original equatorial air is in the region above four kilometres: it has some influence upon what goes on beneath that level, but not enough to justify a claim to identity of air. And yet it must be remembered that by describing the air supply of the warm sector as "equatorial" air the Norwegian meteorologists do contemplate the possibility of flow from equatorial regions. Sometimes in the surface-layers warm air is not truly equatorial, but is deviated from the north-west and warmed by passage over subtropical sea. In the upper air however the continuous flow from the equatorial zone is more probable.

And during the whole period of its travel the air of the revolving column has been subject to solar and terrestrial radiation on the one hand, and its own radiation on the other hand, which would certainly have had some effect upon its temperature and possibly upon its structure. But in any case it will have maintained the habit of arranging itself so that its entropy increases with height, and in that form will be the means of conveying the thermal energy of the equator to the polar regions to be lost there especially by the surface radiation from Greenland and other high lands.

It is certain that air of high potential temperature must pass from the equatorial to the polar regions and air of low potential entropy makes its way from the polar regions towards the equator. That it should be carried in a general current at middle levels in which there are numerous cyclonic circulations, some of which derive themselves from the equatorial region as travelling revolving fluid, is at least not an unreasonable proposition.

25. *The disposal of the evicted air.* I have used the word "eviction" of fluid in the region of origin of revolving fluid without in any way meeting the obvious challenge as to the disposal of the air that has been removed. I have relied simply on the fact that what, according to a graph, is represented by a cyclone is the disappearance of air from the region without compensating increase of pressure in the immediate surroundings.

The disposal of a volume of air of the order of forty thousand cubic kilometres, measured at the surface for a single cyclone, is a serious undertaking. Some of it is represented by the condensed rain, but only about one-eighteenth, or not much more than one per cent., a negligible amount in the rough figures which we are dealing with, though it means 625 million tons of water, about three-quarters of a millimetre of rain over the whole area of 800,000 square kilometres. The arrangement which I should like to imagine is a change of wind-velocity between twelve and fifteen kilometres from the easterly circulation in which the cyclone is formed to a westerly circulation above it, giving a relative velocity of some ten metres per second. In that case the air "evicted" from the easterly current would be delivered into the westerly current and the delivery in the course of a day would mean a block 1000 kilometres long and, allowing for its smaller density at the high level, about 80 square kilometres in section, a sort of smoke trail 1000 kilometres long, 200 kilometres wide and 400 metres thick would dispose of it. Part of it might go to make up the day's loss of 86 metres (at the surface) of the equatorial highs. But what particular anticyclone the debit of any particular cyclone would feed depends upon the special conditions of the circulation.

I have set out these ideas at some length because they suggest lines along which, partly by experiment and partly by observation, some quantitative estimates might be formed of the operation of various physical processes which have hitherto been used in a qualitative manner only in meteorological theory.

It will doubtless occur to the reader that the eviction of air by the frictional forces between the jet of fluid and the air surrounding it is nothing more than the forced blast of a locomotive and is at least as old as Stephenson's "Rocket."

Probably it will also occur to him at the same time that the speed of the jet in such cases is high, and that the action is contingent upon the high speed. Doubtless the effect in the case of forced draught is so contingent, because considerable motive power is required, but there is nothing in the nature of the action between moving air and its environment that limits the evictive action to any particular speed. In the atmosphere the effects are inexorable however slow the relative motion may be, and however long it may take to accomplish them.

*   *   *

26. *Note on the identity of a cyclone during its travels.*

I find that in some quarters these pages have given the impression that I regard the material of which a travelling revolving column of air is composed as remaining unaltered and unalterable throughout its journey, as if it were separated from the atmosphere by an impenetrable wall and conveyed from the place of its birth to the place of its death in something like a huge oil-drum. Such a view is an exaggeration of my opinion. The cyclonic vortex may preserve its identity on its journey and yet change its substance in a way somewhat analogous to that of a passenger on a voyage at the end of which he is the same person but not the same substance. The rotation of the cyclone protects the interior but there is always leakage at the bottom and convexion changes the distribution of mass in the interior and may even be the means of reinforcing the energy.

II. NOTES ON CERTAIN OBJECTIONS TO THE VIEWS EXPRESSED IN THE PAPER ON THE CYCLONIC DEPRESSIONS OF MIDDLE LATITUDES

COMMUNICATED TO THE SUB-COMMITTEE FOR METEOROLOGY, NAVIGATION AND ATMOSPHERIC ELECTRICITY, 25 JANUARY, 1921

THE preceding paper was discussed at a meeting of the Sub-Committee to which it was addressed, and in the course of the discussion a number of considerations were raised tending generally in the direction of discrediting the operation of convexion as a direct dynamical agent in the formation of cyclonic depressions, and of the application of the recognised properties of revolving fluid with a life-history of days or even weeks in the form of a vortex with an approximately vertical core. I was fully prepared to recognise the force of the objections for the special reason, amongst others, that at the time I had not fully realised the possibility for the maintenance of a vortex with vertical axis which is afforded by the resilience of the atmospheric layers, particularly of the stratosphere and other inversions of lapse-rate, and which is more particularly referred to in a subsequent lecture (p. 125).

The considerations raised in the discussion are treated in the following notes.

The effect of falling rain. Attention was called to an error in the decimal point in the calculation of the mechanical action of falling rain as originally written. The printed text of p. 81 has been modified accordingly. As corrected the direct dynamical effect of heavy rain for a minute on the assumptions made (which in any case are not very satisfactory) is very small[1].

It is a matter of common knowledge that a cataract of wind frequently accompanies a heavy downpour of rain. Professor Humphreys, in a recent work on the physics of the air, assigns the cause of the descending wind to the chilling of the atmosphere by the falling rain. Whether the effect is dynamical or thermal the consequence would be to hinder the filling of the vacancy caused by the passage upwards of the air from which the rain is descending and to throw upon the surrounding air near the levels of the raining clouds the duty of maintaining the convexion. The convergence which this action entails upon the horizontal motion would be in effect the same as an annihilation of air at the levels affected and would cause vortical motion there.

The disparity of the horizontal and vertical dimensions of cyclones. Objection was also raised on three other counts against regarding annihilation of air by "eviction" as a real cause of cyclonic circulation. One is the large horizontal area of cyclonic depressions as compared with the vertical dimensions. The relative dimensions may be of the order of 1000 kilometres horizontal to 10 kilometres vertical.

This has been as a matter of fact a favourite objection of my own. I have frequently called attention to the fact that on the scale of the maps of the

---

[1] A calculation of the effect of falling rain has been given to me by Dr S. Fujiwhara and is printed as an appendix, p. 228.

Daily Weather Report the thickness of the effective meteorological atmosphere is not more than ten times the thickness of the paper on which the map is printed, and I have often expressed incredulity as to the air of London being dynamically interested in forces operating in Iceland. There is a sort of proverb at the Meteorological Office that in dealing with a cyclonic circulation we must regard the rotating mass (if it be one) as being in the form of a penny rather than a pencil. I am therefore glad of the opportunity to revert to those opinions and examine them again. No objection really arises on this ground to the development of a small secondary in an established current. So far as the dynamical considerations go any annihilation of air in a column of the atmosphere, even if it were a thousand kilometres away, would of course produce its dynamical effect on the air of London, but as the velocity produced thereby would be inversely as the distance the effects at any considerable distance from the locus of convexion are inappreciable compared with those of more local causes. Certainly one cannot regard a wind approaching gale-force in London as directly and immediately produced by eviction of air in Iceland. But neither need one try to do so. The larger depressions of our latitudes have as a rule a long history. Our local convexions and their consequences are only incidents in that history. The question turns simply upon the energy of the circulation and its disposal. Although the layers affected are very thin the circulation represents an enormous amount of energy for which there is an ample source in rain.

The separation of the loci of eviction and accumulation of air. The immense horizontal extent of cyclonic circulations compared with the vertical dimension renders them especially liable to disturbance by local convexion in two ways. If, as explained later, the air which rises "finds its level" in a locality some 500 kilometres from the position from which it started and produces by eviction a certain amount of annihilation on the way, the locus of eviction may be some hundreds of kilometres from the locus of accumulation or ultimate depository of the evicted air, and the reduction of pressure in the locus of eviction must have as its counterpart an increase of pressure in the locus of accumulation where relative motion ceases, and consequent distortion of the pressure system is produced.

Whether the two loci can be sufficiently concentrated, for convexion lasting for many hours to produce notable effects on the distribution of pressure, depends upon the arrangement of wind-velocity in a vertical section of the original stream. Both effects are inexorable, if they were superposed on the same vertical they would be mutually destructive. Objection was raised that the relative horizontal displacement would be so insignificant in comparison with the vast horizontal dimensions of a cyclonic area that mutual destruction would in fact ensue. I have endeavoured to test the point.

A standard trajectory in the vertical. In order to obtain a representation of a normal current I have made out from the observations of the upper air the normal distribution of pressure at different parts of the earth's surface. And from the normal distribution of pressure at different levels in summer I have computed the normal velocities of the wind in the region of westerlies, i.e.

between latitude 32° and latitude 52°. It gives a curve which starts from a small westerly velocity of 1 m/s at 1 kilometre (probably the smallness is due to the frequent alternation of velocities in the lower layers) and with some pauses attains 20 m/s at 12 kilometres, falling off to zero at 18 kilometres or 20 kilometres. The curve obtained (Fig. 44) is so exactly similar to the typical curves for north and south currents in Cave's *Structure of the Atmosphere* (see Fig. 99) that it can be recognised at once. It is doubtless associated with Egnell's or Clayton's law of the equality of mass-transference at different levels, but that law takes no account of the falling off of velocity at still higher levels, which I regard as essential for the persistence of a travelling vortex.

I have also computed the velocity with which a limited mass of air would rise if its temperature were higher than the temperature of its surroundings, by comparing the velocity with that of a 90-inch pilot balloon as ordinarily adjusted. Assuming that in both cases the law of resistance is purely as the square of the velocity and allowing nothing for the delay of the unenclosed air on account of mixing at its boundary, I find the velocity of the rising air to be a kilometre in 90 minutes, and in the normal atmosphere described the vertical trajectory up to 10 kilometres becomes a curve of which the top is 500 kilometres away from the starting-point. Consequently it is fair to say that the rising air would take advantage of the change of velocity with height to separate its final locus of accumulation from the locus of eviction by hundreds of kilometres. The trajectory with the curve previously mentioned is shown in the accompanying figure (Fig. 44). It is roughly divisible into a nearly vertical part C, a nearly horizontal part A, which must from the nature of the case be much more nearly horizontal on account of slowing down in ascent before stopping, and an intermediate part of moderate slope B. If the supply of convective air at C ceased, the trail would gradually extend itself horizontally at A (because it cannot penetrate the stratosphere) and the result of the convexion would ultimately be a horizontal trail extending over 972 kilometres at the upper level, and there would be a corresponding loss of air distributed over the trajectory.

The loss of air required for the formation of a low-pressure area 200 kilometres in diameter with 10 mb.[1] of depression would be 10^{12} kilogrammes, which would cause an increase of pressure of 2 mb. over a strip 50 kilometres wide along the 1000 kilometres of the horizontal trajectory. Much of the energy would be lost in irregular turbulence and special circumstances would be required for a persistent dynamical system to be formed.

I can hardly contend that in this case the vertical portion is sufficient to cause a definite and persistent circulation and perhaps the effect of the eviction along the line of the trajectory may be only the transient disturbance of a cumulus cloud or a shower, but a more suitable adjustment of the variation of velocity with height is quite within the region of probability, and the line

---

[1] A depression which seemed to form spontaneously over the western front on 1 August, 1917, after four days had a depth of perhaps 10 mb. and a diameter of 750 nautical miles (see Figs. 56–58).

of addition to the mass at the top beneath the stratosphere is certainly inexorable: it must disturb the pressure conditions by increasing the pressure in that region along which it lies.

There are of course endless possibilities of variation in the details of the conditions. I hope to get some further information about the possible

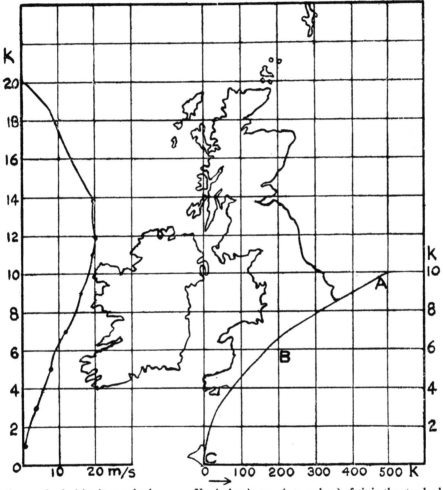

Curve of velocities in standard atmosphere.

Vertical trajectory (o to 10 km.) of air in the standard atmosphere.

The outline of the British Isles is introduced to show the scale.

Fig. 44.

velocity of ascent and typical wind-structures which form the environment; we can thus construct some other examples of trajectories in a vertical sheet. I desired in this one to keep strictly to conditions for which I could give definite evidence and not to select conditions peculiarly favourable for the formation of a vortex.

The wide extent of cyclonic phenomena. The second objection was expressed in the question whether the convexion and consequent reduction of pressure can be supposed to extend over an area large enough to be comparable with that of a large cyclonic depression. I see no difficulty about supposing the process applied in exceptional cases to a column having the area of the core of a tropical hurricane from 10 to 20 kilometres for example in diameter; and all tropical hurricanes are exceptional cases, otherwise there would be many more than there are. When the vortex is once formed, I suppose the convexion at the core itself ceases and is carried on close to the region of maximum velocity surrounding the core.

The ultimate effect of small local circulations upon the major circulation. In the case of our own depressions there are two considerations upon which I hope to get further information. The first is a mathematician's problem as to how far a small vortex created excentrically in a larger vortex may eventuate in increasing the kinetic energy of the original vortex or preventing its decay. I suppose if one had a large number of such secondary vortices the combination would be circulation which would constitute a concentric vortex, but with a single isolated secondary it may be otherwise. The attention of mathematicians might be given to the point[1].

The possible formation of rain areas in cyclonic depressions as the dynamical result of the deviation of surface-wind from isobars. The other consideration is a meteorological problem, namely the extent of the area of rain in our large depressions. Rain certainly represents elevation of air, but not necessarily convexion in the sense of warm air *penetrating a succession of layers* above it, the effect might be produced by gradual and general elevation.

We know that, along the surface, air crosses every isobar from the highest to the lowest at an angle of "incurvature" of about 22° from high to low. In the *Life-History of Surface-Air Currents* Mr Lempfert and I made some attempts to trace the course of the surface air, but the final destination of the air which crosses isobars has never yet been satisfactorily traced. In the idea of the cyclone which used to be accepted the inflow towards the low-pressure centre was regarded as the air which was going to flow upwards. It was regarded as coming down in the adjacent anticyclones, so to speak, for the purpose of providing the material for ascent.

I have shown elsewhere (see p. 77) that so far as the anticyclone is concerned the more appropriate idea is that the air may be regarded as taken from all over the area covered by the closed isobars of high pressure, and represents not a descending current but a slow settlement of the whole mass. For the great Atlantic anticyclone I estimated the rate of settlement at 86 metres a day, other examples of smaller anticyclones have given three or five times as much.

It is possible that the area of the cyclone may present the counterpart, that is to say that the air which flows across isobars towards the centre may not become a local rising current but an increasing accumulation which pushes

---

[1] Since this was written Dr S. Fujiwhara has presented to me a paper, not yet published, which deals with the growth and decay of vortices through their action on their environment, and the analogy with other cases of growth and decay.

up the whole superstructure of the cyclone. In that case we may get rainfall from the gradual elevation of the cyclonic mass over a large area with possibly local convexion in addition. In the *Life-History* Mr Lempfert and I endeavoured to account for rainfall by using trajectories to identify localities of convexion. The process was not very satisfying and we did not deal with the general elevation. Yet such elevation is wanted to account for the conditions of temperature in a cyclone just as the corresponding settlement is needed for the conditions of temperature of anticyclones.

I have pointed out elsewhere that with a uniform angle of incurvature or outcurvature the outflow will represent a uniformly distributed subsidence in any region where the velocity is proportional to the radius (v/r = constant), or, as we may express it in a manner suitable for isobars of irregular shape, where the velocity is inversely proportional to the curvature[1]. On the other hand where the velocity is uniform for successive straight isobars or where the velocity is proportional to the curvature (vr = constant) no ascent or descent is required; the air simply passes across the area without loss or gain. Corresponding statements are equally true for the convergence of a cyclone. Making some allowance for exposure, which is not easy to adjust, the inflow to a cyclonic area can be ascertained and an estimate formed of the amount of elevation caused thereby in different parts. It is possible that the accumulation of air may reach such dimensions as to justify the formation of rainfall by that cause. It is to be noted that this is essentially a *dynamical effect*, the inexorable result of the juxtaposition of high and low pressure areas and the friction of the ground, which prevents the air near the ground from settling itself in a steady state of motion in relation to the isobars.

The rainfall which results from this convergence is the cyclone's reaction to the threat of increasing pressure at its base which otherwise would "fill it up" more quickly than is the actual case. So even in this case, but for a special reason, the rainfall maintains the energy of the cyclone.

The formation of low pressure. The third objection was expressed by the words "actually, of course, if some of the ascended air is carried away its place must be taken by other air and in order that the mass of air in the given vertical may be decreased the replacing air must be the lighter. If this is so it would seem impossible for the ascending air to rise." The description does not represent the events as I picture them. I take it that for any layer the disappearance upwards of any air potentially warmer than its environment is inevitable. The process may be swift or slow, simple or complicated, by threads or by "blobs" or any other form, but anyway it is inexorable. Ultimately the warmer has got to leave its potentially colder environment. The process of forming reduced pressure is not dependent on the density of the air which replaces that which has gone, but upon the fact that the initial driving force comes from the sides and is expressed by convergence; the convergence causes circulation which in its turn protects the low pressure.

---

[1] Roughly the relation between vertical velocity v within an area and the wind-velocity V of an isobar surrounding it, with incurvature $30°$ is $v = \frac{1}{2}Vh/r$ where h is the height over which the inflow is distributed.

If the area of low pressure takes more warm air from below the process goes on until the base is protected by a floor of air with entropy diminishing downward, and, in the region above, the invasion from the sides goes on until permanence is assured by reaching a ceiling where there is increase of entropy upwards. If necessary the stratosphere will become the ceiling. It has the necessary resilience. The resilience of the ground is provided by its mass; of water likewise; of the stratosphere by its entropy. If these protections cannot be attained the reduction of pressure is transient and there are doubtless many unfavourable cases in which that is the result, but not all cases are necessarily unfavourable.

Objection is also taken to my suggesting that stability or, I should prefer to say, persistence is secured by the vortex "being gradually extended upwards with diminished rotation." But this is certainly possible when the vortex is extended to the stratosphere where a ceiling of isothermal air will afford the necessary protection.

I do not think there is any satisfactory evidence from actual observation of circulations or apparent circulations being persistent for days, as our depressions are, unless the stratosphere is brought into operation. I suppose that in the case of "dust devils" and other temporary whirls persistence may be secured for minutes, or even for an hour or two, by the vigour of the vertical convexion gradually extending upwards, if my estimate of 90 minutes for the traverse of a kilometre is reasonable; but for the extension of the persistence from minutes to days there is so far as I can see no alternative to the employment of the special properties of the stratosphere.

The importance of the frictional forces. An opinion was put forward with the apparent concurrence of the mathematicians present that if the frictional forces between rising air and its environment were of importance in the eviction of air from the environment, in the course of the convexion, similar forces would be equally important in destroying the circulation of the air round the column of low pressure; and the combined effect would therefore necessarily be zero in all cases, much as though it were positive and negative electricity that we were dealing with. This seems to me to raise a fundamental question with regard to the dynamics of the atmosphere which mathematical physicists are much better able to solve than I am. I have always supposed that if the motions of the atmosphere were so thoroughly irregular, as say the motions of molecules in a gas or the complexity of eddies in wind near the surface of the earth, the energy would be speedily dissipated; but if the motions could be arranged and organised so that the relative motion of the parts was reduced the persistence of the systematic motion would be of an entirely different order from that of the casual and irregular motions. To this general principle I have been accustomed to attribute the transient persistence, such as it is, of the eddies at street corners. The rotational system is stable with the ordinary irregularities. There is a sorting effect more or less analogous to that which Maxwell contemplated as affording a possible exception to the second law of thermodynamics. The dynamical effect of the rising air upon its environment is of the nature of irregular eddy-motion; there is everywhere a good deal of discontinuity whereas the motion of air in a vortex is of the systematic kind in which the variation of velocity is continuous and very gradual; and, in consequence, the importance of the frictional forces in destroying motion is comparatively small. The objection could be applied with equal force in the case of the vortex ring and it would be just as legitimate to say that the frictional forces which are of importance in forming a vortex ring are equally important in destroying it, and therefore the result of the process of forcing air through an opening is naught so far as vortex rings are concerned. There are so many other cases which illustrate the advantage of systematic arrangement in conserving energy from loss by friction that I feel sure a distinction may fairly be drawn between the frictional effects of air forced through other air by gravity and the organised rotation of an atmospheric vortex.

The spontaneous formation of cyclones. Primary objection was also raised on the ground of experience to allowing convexion as a cause of formation of cyclonic depressions of substantial magnitude, particularly to the application of convexion to explain the formation of tropical revolving storms, as equally to the direct application of convexion over the heated continent of Asia to explain the monsoons. I agree as to the monsoon winds. I suppose that the rains of the monsoons are really due directly to convexion but the general winds are not; and I think we may consider the effect of convexion upon a main air-current of the atmosphere quite apart from the question of the origin of the main current itself.

The objection was directed in particular against considering convexion as a cause of cyclonic storms in the Bay of Bengal which Blanford and Eliot had particularly cited as examples of that effect. The view was put forward that the energy of local convexion is dissipated in a large number of effects which are distributed sporadically over a large area and have no co-ordination. On that view an inflow of cold air into a hot environment would be regarded as probably the originating cause of the cyclones. The effect of convexion would be restricted to small disturbances like cumulus clouds or dust-devils; the larger examples would be excluded from its direct influence. The formation of vortex whirls in the smoke of a cigarette when it was brought into relation with a cold horizontal surface was cited as an example of the action.

This again raises the very large general question of the combined ultimate influence of a very large number of small vortices. The case of the air over the sea is different from that over the land because the sea-air contains a large quantity of moisture which potentially increases its entropy and ultimately its potential temperature, and therefore indicates a very high level as its ultimate locus of accumulation. We have therefore to consider not merely the convexion at the surface, where there is little general motion of the air, but also the subsequent convexion at all levels up to 15 kilometres within which there may be air-currents of very considerable intensity. Meanwhile I would remark first that there seems to be practical continuity in the sizes of the various examples of rotational systems from the cigarette smoke to the circumpolar whirls of the two hemispheres, and secondly that a sufficient number of small circulations at the surface would integrate kinematically to give a circulation round the enclosing boundary[1].

For the furtherance of the consideration of this subject an investigation of the upper air over the tropical oceans would be very helpful. We have practically only investigations over the eastern Atlantic and that is a region where cyclones are at most very infrequent.

The amassing of cold air as a substitute for convexion. From the fact that at all levels in the troposphere cyclonic regions are cold in comparison with anticyclonic regions it was suggested, as a kind of reasoning from the converse, that the coldness may in itself constitute the cause of the cyclone irrespective of convexion, perhaps with the implication that the line of demarcation between hot and cold in a cyclone has no symmetry at any level. It certainly seems easier to explain how convexion produces a cold-cored cyclone than how cold air could concentrate in a column or surround itself with warmer material, but the suggestion is arresting because it throws some light on the question of the relation of pressure to temperature in the upper layers. Clearly if the air is cold where pressure is high the pressure-difference will be diminished with height and will be reversed if the temperature-difference is consistent for a sufficient height. Hence a warm anticyclone and relatively cold cyclone form the only arrangement which can be consistent through any considerable range in the vertical. And hence one ought not to be surprised that in the higher layers (4 km. to 8 km.) pressure and temperature are closely correlated. One could not otherwise get the structures suitably ended off at the top, and so structures which are not in accordance with that arrangement are probably transient.

The essential importance of convexion. I have recalled these objections and have answered them because, when all is said and done, thermal convexion certainly exists and is the only original dynamical agent which is available for meteorological theory. Everything must ultimately be referred to it. Looking through Prof. Humphreys' recent work on the *Physics of the Air* I find nothing else suggested as a possible dynamical agent except volcanic eruptions, which are referred to as throwing dust into the stratosphere apparently by direct dynamical action. But one can hardly regard the intermittent operations of volcanoes as responsible for the multifarious phenomena of weather. The direct dynamical influence of the heavenly bodies is no longer regarded as sufficient. The winds affect the waves and currents of the sea, but the direct dynamical effect of moving water upon air is trifling, and in fact the motion of the water may rather be regarded as dependent on the motion of the air.

If a cold current of air penetrating a warm environment is required for the

---

[1] It was remarked by a mathematician present that the mathematical treatment of a number of vortices in the same field was difficult, and that the natural process was, probably, for a single vortex to break up into two or more, rather than the reverse.

dynamics of a cyclone, or a cold pile of air is required for the origin of a depression, we must go ultimately to thermal convexion to provide them. There is nothing but thermal convexion to act as the motive power for every drop of rain that ever fell, and for every wind that ever filled a sail or wrecked a ship since the world began.

The situation seems curious; we are all agreed that in the aggregate convexion is responsible for all the winds that blow. We recognise that the convexion of the present operates not upon still air but upon the existing circulation with its momentum due to the convexion of the past; yet when we come to actual individual winds as we find them on the planet the tendency is to "turn down" as ineffective and inapplicable the only agency we have, or can have, for explaining their ultimate origin. If we do so turn it down we should have to accept the position that convexion having first dissipated all its energy in dust-devils and other minor frivolities, somehow or other manages to pull itself together out of its dissipation and conduct a remarkably active business in winds—trade-winds, monsoons, cyclones, tornadoes, cyclonic winds and anticyclonic winds—all over the world. I am disinclined to accept that as an accurate view of the position. I do not think it necessary to allow that convexion begins by dissipating all its energy and indeed it would not help matters much if it did, because its dissipation only means a development of heat in a different locality and the process of convexion would have to begin all over again.

I would rather take the position that convexion is not so dissipated as we have been inclined to think, that it has ways of conserving the kinetic energy that it generates, different perhaps from those which our predecessors have imagined, but still in agreement with the ordinary laws of dynamics and physics. I still think that one of the direct effects of convexion in a broad current of air is a reduction of pressure in some part of the track of the rising air; there is a connection between that inexorable physical fact and every wind that blows on the globe, and we ought to be able to trace the connection if we are careful to pay stricter attention to the operative conditions than the early meteorologists did, and have the mechanical skill to analyse the complicated operations. It seems useless to suggest that convexion is not operative; precisely how it operates is a fair subject of inquiry.

SYMMETRY IN THE HORIZONTAL DISTRIBUTION OF TEMPERATURE OF THE
AIR IN CIRCULAR ISOBARS OF MIDDLE LATITUDES BETWEEN
4000 M. AND 8000 M.

The correlation coefficients contained in the table on p. 85 mean *primâ facie* that there is a relation between changes of temperature in the horizontal at all levels between 4000 and 8000 metres and the corresponding changes of pressure that approaches very nearly to direct proportionality. The result of Mr Dines's observations and calculations, which is summarised in this way, falls in very well with the idea that a horizontal section of a cyclone between those levels may be regarded as a circulation round a centre with a distribution of velocity according with the gradient of pressure. The centre may, of course, be a moving centre carried, along with its proper circulation, in a more general current and the natural conclusion from proportionality of pressure-changes to temperature-changes is that in a cyclone with a well-marked centre and circular isobars, isobaric lines are also isothermal lines, temperature is not dependent on the direction in which the air is moving but is as symmetrical with regard to the centre as pressure is; and if the proportionality held rigorously the conclusion would be extended beyond the limits of a circular cyclone to other types of barometric distribution. Since the observations in Mr Dines's collection are taken at random and the coefficients of correlation are very high, there is *primâ facie* evidence for that state of things. Such a general *régime* for the layer of air between four and eight kilometres in our latitudes is in marked contrast with experience at the surface and could be used as an argument against the existence of any notable discontinuities of temperature at those levels and in so far is evidence against any marked surface of separation between polar and equatorial air in those layers.

This inference from the observations of the upper air has been adversely criticised in some quarters. The objections are based partly upon the general ground that when two quantities are correlated with a coefficient r the fraction of the one quantity which is controlled by the other is r^2, and hence with a correlation coefficient ·7 only one half of the variation of the correlated quantities is controlled by the other, leaving another half to be controlled by other causes. This kind of reasoning is not very cogent in the case of the observations of the upper air because the coefficients in that case are higher than ·7 and are indeed, without any allowance whatever for the errors of observation, of the order of ·8 or ·9, and we may conclude that three-quarters of the variation of temperature is due to variation of pressure. Very little is left for the operation of any single one of the many other causes which may be supposed to affect the temperature of the air, source of supply, convexion or radiation, for example. Another objection taken is that discontinuities of temperature, such as would confirm the suggestion of a polar front at higher levels, have been actually noted at four kilometres or higher, by observers in aeroplanes, and occasions have also been noted when increased temperature was accompanied by reduced pressure at the same level, or *vice versâ*.

Further it has been argued that a high degree of correlation is natural as a general consequence of the dynamical conditions. That, of course, must be the case if the correlation exists; the remarkable thing, however, is that the correlation should be so extraordinarily high. If proportionality were a universal rule for the levels mentioned, isobars would be isotherms everywhere within those levels and therefore the mean isobars would be parallel to the mean isotherms. Such information as we can gather from the figures for England seems to indicate that that is the case; but we have not sufficient observations to enable us to test the point effectively for variations in latitude and longitude. It remains, however, an important question for investigation.

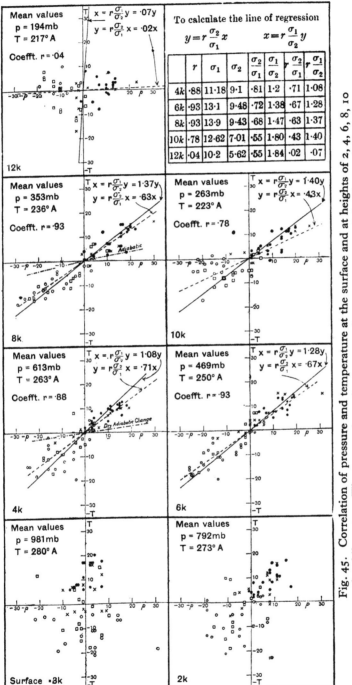

Fig. 45. Correlation of pressure and temperature at the surface and at heights of 2, 4, 6, 8, 10 and 12 kilometres. The mean values of pressure and temperature at each level are taken as axes.

□ Winter (Dec. Jan. Feb.)   □ Spring (March, April, May)   • Summer (June, July, Aug.)   × Autumn (Sept. Oct. Nov.)

Hence it follows that the best way of exploring the meaning of the extraordinarily high values of Mr Dines's coefficients is not to discredit the figures or ignore them, but to examine the corresponding figures for other regions.

For this purpose Miss E. E. Austin has taken out the pressures and temperatures obtained recently at Woodstock, Ontario, by Mr J. Patterson for the Canadian Meteorological Office. They are a very homogeneous series obtained by apparatus which Mr Dines supplied. The results are set out, pressure and temperature as co-ordinates, in diagrams for the surface, 2000 metres, 4000 metres, and so on up to 12,000 metres (Fig. 45). A separate symbol is used for each of the four seasons. The diagrams speak for themselves, the correlation coefficients and standard deviations for the whole year are given and the regression lines are drawn. There are not enough observations to justify the same treatment of each quarter. The manner in which the two regression lines close up between four and eight kilometres and open out again above and below is quite remarkable and exhibits the distinctive character of those layers.

In Canada clearly the seasonal influence is paramount. The summer is a period of high pressure and high temperature, the winter of low temperature and low pressure. But for such observations as there are the coefficients for the several seasons are high, for example, for four winter months (Nov., Dec., Jan., Feb.) the correlation at 4 km. is ·8 and at 8 km. ·9; for four summer months (May, June, July, Aug.) ·7 and ·8 respectively.

Among the Canadian observations for consecutive days we have traced one case in which a decrease of pressure was accompanied by an increase of temperature in winter. It was due to a surge of equatorial air replacing one of polar air with a transition from the pressure-temperature relation of the north to that of the south. Mr Dines's figures indicate that this country is less subject to that kind of transition than Canada.

The remarkable fact about this change of temperature incidental to change of pressure is the large change of temperature which results. It is several times that which corresponds to dry adiabatic change which is indicated in Miss Austin's diagrams by a chain line. For England at the level of eight kilometres the change is the adiabatic change itself; below that level, and above it, it is much greater. Mr Dines has expressed these facts by supposing the air drawn out from low pressure to high pressure along the level of eight kilometres, that is just under the stratosphere of a cyclonic area, and drawn upward and downward from below and above in the cyclonic area, and pushed downward and upward in the anticyclonic area.

III. EXPERIMENTAL ILLUSTRATIONS OF THE "EVICTION" OF AIR

PAPER COMMUNICATED TO THE SUB-COMMITTEE FOR METEOROLOGY, NAVIGATION AND ATMOSPHERIC ELECTRICITY, JUNE, 1921

IN a previous communication[1] I have suggested that the removal of air along an axis, which was postulated by the late Lord Rayleigh as a stage in the development of a simple vortical circulation about the axis, might be accomplished in the atmosphere as the indirect effect of convexion regarded as localised upward motion of air. In the free atmosphere the original cause of the upward motion might be statical instability such as would result, for example, if air with a dew-point of 300 a were slightly warmed at the surface of the equatorial region, while it had the environment represented by the normal condition of the atmosphere at Batavia (lat. 6° S.) up to 20 kilometres. The average surface temperature is 300 a and the lapse-rate for that region is slightly greater than the lapse-rate of the adiabatic for air saturated at 300 a.

In the hypothetical case, therefore, the air warmed at the surface would rise, and in doing so would pass through the series of changes represented by the isentropic or adiabatic line for air saturated under normal pressure at 300 a. It would have to penetrate to the level of 15 kilometres before it found a position of stable equilibrium and would therefore cross the strata of air between that level and the ground with some finite velocity. The measure of the velocity at any level is at present unknown, but judging by the lifting of hail-stones it may in favourable circumstances be very great. My suggestion was that on the journey in question it was not merely the original air that made the ascent but also all the air of the environment that became entangled with the original rising air by eddy-motion. The whole would be carried upward with a suitable reduction in the velocity of vertical motion developed by the convexion.

The question of how much upward velocity is necessary to cause eviction obviously arises in this connection. I suggested therefore that the process might be illustrated experimentally by causing convexion in a vertical column provided that arrangement could be made for giving some initial relative motion to the different parts of the air which was induced by the dynamical action to approach the core, and also that the natural tendency of the free atmosphere towards stratification in horizontal layers could be imitated. I have constructed an apparatus which appears to me to satisfy the conditions and I propose to demonstrate its action at the meeting on June 21. The tendency to stratification in the atmosphere may fairly be regarded as due to the deviation of the actual condition of the successive layers from those of

[1] "The Cyclonic Depressions of Middle Latitudes," *Met.* 76 (see p. 80).

convective equilibrium, and that again may be expressed as increasing potential temperature, or realised entropy, of successive layers from the ground
upwards.

Any deviation of the distribution of temperature towards stability from
that of convective equilibrium leads to forces of restitution for vertical displacement, which would not exist in the case of an atmosphere in convective
equilibrium or in an incompressible liquid of uniform temperature; and thus
the deviation of the several layers of the atmosphere from the condition of
temperature required for convective equilibrium endows the layers with a
species of resilience to. vertical displacement which is quite analogous to the
resilience of a surface of water for upward displacement, but which, estimated
numerically, is very much smaller. Thus the resilience of the stratosphere
which is approximately isothermal in the vertical, is of the order of one ten-
thousandth part of the resilience of a horizontal surface of water, because a
depression of the tropopause through a kilometre is necessary to balance a
pressure-difference of 10 mb., which can equally well be balanced by an
elevation of 10 cm. of water. A layer comprising an inversion of lapse-rate
would be more resilient than the stratosphere; but a layer with positive lapse-
rate would be less so, until with an adiabatic lapse-rate and a state of neutral
equilibrium, there would be no resilience at all.

In any case, the displacements necessary to counterbalance a millibar of
pressure would be of the order of a hundred metres, and therefore this
property cannot be brought into play in a laboratory experiment. I have
supposed that to imitate the conditions in a laboratory experiment it would
be sufficient to constrain the stratification by horizontal plates perforated
with a circular aperture intended to leave a free passage for the column of air
with vertical motion; but to secure that the air moving to replace that which
has been evicted by the rising column should approach the core with a motion
approximately horizontal. As a first experiment the circular apertures in the
glass plates have been given a diameter of two inches. At the same time the
initial relative motion is obtained by giving a tangential-motion to the
approaching air. For this purpose access to the core is limited by the cells
of a "wheel" made of two parallel flat metal rings, the space between the
two rings being separated by a series of thin vertical partitions of wood nearly
tangential to the inner circle, forming a sort of honeycomb with oblique
partitions, the cells of which give a tangential component to the air entering
the circular enclosures, between the glass plates. The glass plates are 50 cm.
square and the honeycomb wheels have an external diameter of 47 cm. and
an internal diameter of 37 cm.

In the first pair of honeycomb wheels the separation of the circular plates
was fixed at 1·25 cm. Calling this axial dimension the *thickness* of the wheel
a third wheel 2·5 cm. thick was constructed and a fourth of 5 cm. thick. The
wheels can be laid one on top of the other to form a single cylindrical pile
10 cm. in height. It appears that there is no advantage in having very thin
wheels; 5 cm. is quite sufficiently thin.

The chamber formed by the superposed wheels can be divided horizontally by laying a perforated glass plate on the top of each separate wheel.

With the arrangement thus provided for the admission of air a vortical whirl can easily be created by extracting air from the top, if the opening in the top plate be covered by a metal chimney 7·5 cm. in diameter and the chimney operated by a gas burner. Air can in that way be extracted from the chamber through the 5 cm. circular orifice, but this method of removing air from the core of the vortex does not differ essentially from the mode of extraction by means of a fan-wheel employed by Mr W. H. Dines in the model in the Science Museum. My purpose was to restrict the experiments to cases in which the first stage of the operations was to deliver air or other material *into* the enclosure with an upward motion, and so imitate the natural

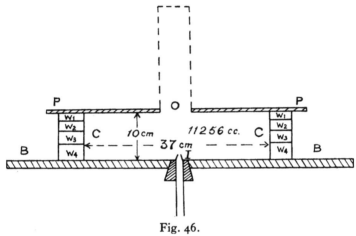

Fig. 46.

BB is the base board with the operating circular aperture I forming the inlet. PP is the covering glass plate with the outlet orifice O. W_1W_1, W_2W_2, W_3W_3, W_4W_4 are the honeycomb wheels which surround the chamber CC and give access to it, through the honeycombs, for any air that may be drawn in by the action of the operating orifice I.

case of convexion in which the working material operates from below and not entirely from above.

In order to operate in this way the chamber was set up on a board which was perforated with a circular hole 5 cm., or rather more, in diameter, and thus an apparatus was formed of which a vertical cross-section was as shown in the accompanying sketch (Fig. 46).

The indication of the motion of the air in the chamber CC was obtained by smoke or by small balanced wind-vanes.

My first object was to ascertain whether eviction could be secured by injecting air in the form of a stream directed vertically through the inlet I to the outlet O, and for that purpose I used a powerful jet through an inch nozzle from the motor-driven fan of a vacuum cleaner. A vortex was produced immediately and was quite unnecessarily vigorous. The jet rising through the outlet aperture O "tore the heart out" of the axial portion of the chamber. It carried thread and other light material to the ceiling of my laboratory. The smoke made less than one complete revolution within the wall of the chamber before it reached the core and vanished in the rising current: the small vanes showed an angle of incurvature of about 45°.

I then reduced the blast, replacing the electric blower by foot-bellows which fed air into the chamber through a glass tube about 7 mm. in diameter. The glass tube was held in a perforated cork which filled the inlet aperture of the chamber. With that arrangement a vigorous vortex was formed almost instantaneously round the jet of rising air and maintained as long as the blast continued. The action still seemed unnecessarily strong, but having no means available at the time for working with less power I turned my attention in another direction and used the apparatus to repeat Aitken's experiment on the formation of a "secondary" in a rotating liquid by falling sand, with air instead of water. My apparatus satisfied Aitken's conditions with a fixed position for the axis of the created vortex instead of a travelling one. The initial circulation or relative motion, which Aitken provided by the rotation of the liquid, was secured by the shape of the inlets in the honeycomb wheels leading to the chamber. I could therefore work with a fixed reservoir for the sand instead of carrying the supply round with the fluid.

To obtain this modification I removed the cork and left the "inlet" orifice unobstructed. In a stand on the glass cover of the apparatus I mounted a funnel with its nozzle about a foot above the outlet orifice O and poured sand into the funnel. The result was the reverse of that of the air-jet, air was carried into the chamber by the falling sand and the smoke was blown outwards from the chamber through the honeycomb walls by the air which the falling sand brought into it. I may remark that the air issuing in this way showed no tendency to form a persistent whirl corresponding with an anticyclone, though the original tangential motion and the customary divergences were provided. Such conditions are apparently not sufficient for imitating an anticyclone. I have not yet ascertained what additional conditions are required. In order to produce extraction instead of compression I replaced the funnel by a cardboard cylinder with perforated base 65 mm. in diameter, which stood on the glass cover itself and surrounded the inlet orifice. When supplied with sand it provided half a dozen streams of falling sand which started from the cover, passed through the chamber and were caught in a bucket beneath it. With this arrangement there were distinct traces of vortical motion developed in the chamber but not enough for satisfactory investigation.

As I wanted more air removed and had no more honeycomb wheels with which to build up a chamber of greater depth than four inches I fitted a metal tube into the lower orifice and so got the advantage of the effect of the falling sand upon a column of about 80 cm. in height. This gave a very satisfactory vortex in the chamber which showed the smoke making more than a complete revolution in the chamber before it was carried down by the falling sand. The accumulation of smoke in the bucket with the fallen sand is very striking.

The sand fell at the rate of 1·4 kilogrammes per minute, but the vortex was fully established in a few seconds. I used about half a litre of sand (sp. gr. 2·6). A steady state was reached almost instantaneously.

It was then time to get some measurements and for that purpose the method of the air-jet seemed the most easily applicable. Accordingly I restored the air-jet, but arranged to feed the supply of air by a pair of aspirating bottles instead of bellows. Water running through an indiarubber tube from the bottom of one aspirator bottle to the bottom of another, which was placed below the chamber and had its top connected directly with the tube above it, pushed the air into the chamber by simple displacement. The aspirating bottles are rated to hold five litres of water, allowing for the height of the tubes of communication in my arrangement, about four litres of air are expelled from the second bottle through the chamber while the water runs from the first bottle to the second after it has been lifted onto a suitable stand. The water runs back again when the first bottle is lowered to the ground and the four litres of air pass back again into the first bottle. Thus the effects of direct suction from the enclosure and the induced suction, due to the degradation of the kinetic energy of the same amount of air forming an issuing jet by eddy-motion between it and its environment, can be compared.

With the stand that I had available the difference of water level at the start of the flow was 110 cm. and at the end 70 cm. It took 30 seconds for the four litres of water to flow between the two.

With this arrangement, as with the sand, the direct passage of the air of the jet through the 10 cm. of the chamber to the open air beyond the outlet orifice gave some indication of the formation of a vortex by the extraction of air from its core, but so slight as to be easily disturbed by the casual air-currents in the room outside the chamber. The effect of extraction of four litres of air by direct aspiration through the bottom into the aspirator was of the same kind and unsatisfying. The effect of the jet was therefore reinforced by surrounding the outlet orifice with a cylindrical tube of cardboard 6·3 cm. in diameter and 30·5 cm. long standing on the glass cover. With that modification an excellent vortex was set up in a few seconds and maintained throughout the period of the flow. It continued also for an appreciable time after the conclusion of the flow with a gradual expansion of the curve of the spiral.

The amount of air delivered in this way was measured by an air-meter, which consists of a fan-wheel with counter in a circle of the same diameter as the reinforcing tube. The total linear flow along the tube for one discharge of the aspirating bottles was 50 feet in half a minute, allowing nothing for correction on account of friction. When the reinforcing tube was removed and the air-meter covered the outlet opening it registered 40 feet in half a minute, and when the head was reduced by using a lower platform for the driving aspirator the flow was 34 feet per half minute. In that case the initial head for flow was 16½ inches and the final head ½ inch.

The figures given indicate the discharges produced by the passage of four litres of air through the nozzle and correspond with motion past the air-meter (uncorrected for friction) of

0·5 m/s for head 110 to 70 cm. and reinforcing tube,
total delivery 45 litres in half a minute.
0·4 m/s for head 110 to 70 cm. without reinforcing tube,
total delivery 36 litres in half a minute.
0·33 m/s for head 41 to 1 cm. and reinforcing tube,
total delivery 30 litres in half a minute.

The total deliveries are surprisingly large; the air used to operate the machine measured four litres in half a minute. Hence the four litres by the degradation of its initial kinetic energy caused the eviction of 41 litres from the chamber in the first case, 32 litres in the second and 26 in the third.

Thus the air *evicted* amounted to ten times the amount of air which produced the eviction in the first case, eight times in the second, and six-and-a-half times in the third. The volume of the chamber within the honeycomb circles is 11·25 litres. Hence by one discharge of the aspirators the chamber would be evacuated $3\frac{1}{2}$ times, 3 times, and $2\frac{1}{2}$ times respectively. This conclusion can be roughly tested by watching the smoke during the operation. It does not appear to be an exaggeration of the power of eviction.

And, if it is not, the result is reassuring because not much additional air, requiring to be disposed of somehow in the layers under the tropopause, would have to be included with the air evicted, in order to form a cyclonic vortex. Thus, the amount of "surface" air estimated as having been evicted in the Cocos Island cyclone is 40,000 cubic kilometres; only 4000 to 7000 kilometres need be added to represent the evicting air on the scale of relative magnitude indicated by these experiments. The vortex which is produced by sand is of like intensity with that produced by the air-jet and similar amounts of air are presumably removed. *Primâ facie* objection might be taken to the use of a reinforcing tube, either for sand or air; but the objection is not serious. The eviction takes place; the only question is as to its distribution in the vertical. That question could be pursued to a solution by multiplying the number of honeycomb wheels, thus increasing the height of the chamber. Some interesting experiments could then be made as to the outflow of the air at the top in the case of the air-injector and at the bottom in the case of the falling sand. From them we might learn something further about the idea of an anticyclone as a region of descending air.

The apparatus was constructed with the assistance of Dr J. S. Owens, and our intention was to use thermal convexion as the motive power for the eviction. I have tried the effect of closing the inlet aperture with the bulb of an incandescent lamp; but the convexion produced thereby is not vigorous enough to display the formation of a vortex. Further experiments with that object were in progress at the time. They showed that vortical motion resulted from the eviction caused by an incandescent lamp surmounted by a funnel which led to the bottom of the chamber.

8. STRUCTURE OF THE ATMOSPHERE UP TO TWENTY KILOMETRES

GEOPHYSICAL DISCUSSION, ROYAL ASTRONOMICAL SOCIETY, 6 MAY, 1921

THE picture that I have had vaguely in mind for many years is that of a vast heat-engine, or rather of a collection of engines, working according to thermo-dynamic laws, of which at any time the working substance is part of the air while the remainder of the atmosphere forms the environment or cylinder. I should like to lead up to that picture. I propose, therefore, to sketch the outlines of it as I see them, and afterwards to fill in the details as far as I can. My reason for adopting this plan in the opening of a geophysical discussion of the structure of the atmosphere is that, in practice, this is one of the rare occasions when the physical aspect is to be kept in view as distinguished from the more purely dynamical aspects. As a rule contributions to the physics of the atmosphere have been of the nature of marginal notes to a text which exists, if at all, only in the imagination of the several authors. I have put my own mental picture of the physical conditions into black and white. It may turn out to be wrong either in principle or in detail. If my picture is correct we can use it effectively as a background of reference for considering the physical processes which are suggested for the atmosphere.

Let us first take a general view quite irrespective of details and of any question of explanation of the phenomena.

1. STRATOSPHERE, TROPOSPHERE AND TROPOPAUSE

It is now common knowledge that the atmosphere between the surface and 20 kilometres is separated into two parts which have been called the *strato-sphere* and the *troposphere* respectively. The existence of a definite surface of separation between the two parts was brought to light by determinations of temperature and pressure obtained by means of sounding-balloons carrying self-recording instruments.

The stratosphere is the upper part: in it the isothermal surfaces which represent the distribution of temperature are approximately vertical. The lower part is the troposphere in which the surfaces of equal temperature are approximately horizontal. The line of demarcation between the two where the troposphere ends and the stratosphere begins has been called the *tropopause*: it is found at heights between 7500 and 17,000 metres, depending upon latitude and conditions of weather. The available observations show that the tropopause slopes upward towards the equator, downward from the equator polewards. It also slopes downward locally towards the centres of cyclonic depressions and upward towards maxima of pressure. The localities thereby indicated as regions of lowest stratosphere are in both cases, though for

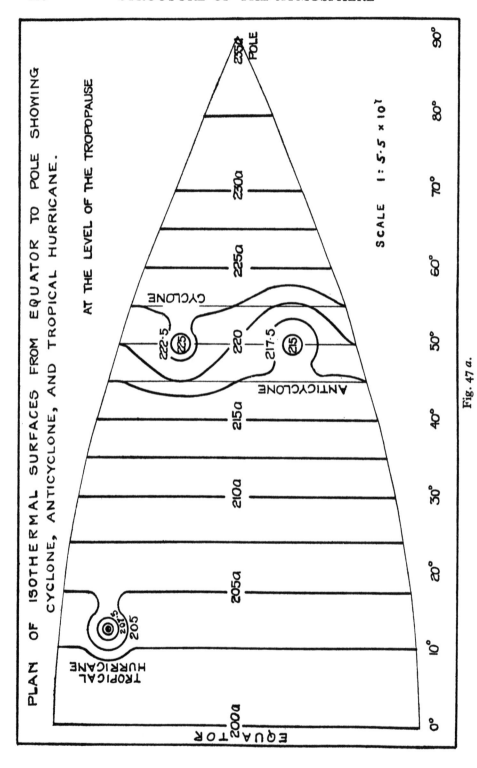

PLAN OF ISOTHERMAL SURFACES FROM EQUATOR TO POLE SHOWING CYCLONE, ANTICYCLONE, AND TROPICAL HURRICANE.

AT THE LEVEL OF THE TROPOPAUSE

SCALE 1 : 5·5 × 10⁷

Fig. 47 a.

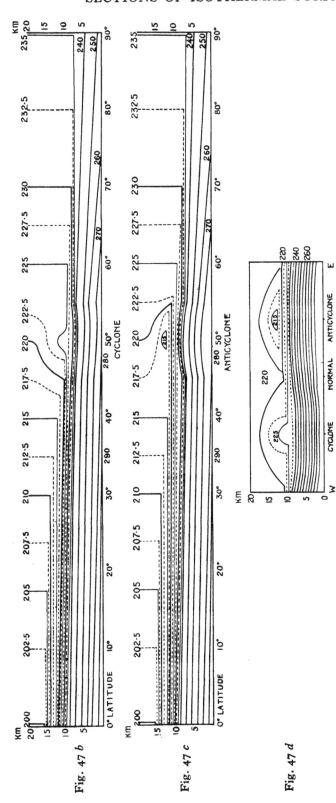

Fig. 47 b.

Fig. 47 c.

Fig. 47 d.

Fig. 47 b. Vertical section of a model of the isothermal surfaces in the atmosphere up to 20 kilometres from equator to pole passing through the centre of a cyclone in latitude 50° N.

Fig. 47 c. Vertical section of the model from equator to pole passing through the centre of an anticyclone in latitude 50° N.
Fig. 47 d. Vertical section of the model from west to east along latitude 50° N. passing through the centres of the cyclone and anticyclone.

Temperatures are given on the tercentesimal scale.

Note. So far as possible the model is based upon the recorded observations in the upper air, but many details are lacking, and until they are available the details of the model are hypothetical. The reader should compare the hypothetical structure through the anticyclone (*c*) with the ascertained structure on 29 July, 1908, shown in Fig. 66.

different reasons, very unfavourable for observations. The polar regions are uninhabitable, or at least inhospitable, and the neighbourhood of centres of cyclones are peculiarly stormy and variable. For that reason we can assign no definite value for the lowest level of the tropopause. The highest level is fairly well-defined by observations in the equatorial region, which is characterised by constancy rather than variability. We are not quite sure whether the tropical anticyclones over land would give slightly higher values than the equator, but the differences cannot be great.

2. The Continual Mixing of the Air

We find the air of the troposphere to be persistently and thoroughly mixed so that the proportions of oxygen, nitrogen, carbonic acid, argon and other permanent gases are invariable. The processes of mixing are less obvious in the stratosphere; but so far as we know the composition does not vary and probably mixing, though much less vigorous than in the troposphere, is effective up to the level of 20 kilometres. The dust of the Krakatoa eruption gradually diffused itself at great elevations over both hemispheres and other volcanic eruptions have been followed by pale skies and other signs of fine dust at great heights over vast areas. Direct observations up to 14,000 metres were made by Teisserenc de Bort, who devised an ingenious apparatus to capture specimens of air at those levels for the purpose of analysis in the hope that they would show larger proportions of the rarer gases. But on examination, with Sir W. Ramsay's assistance, they proved to be common air.

3. Peculiarities of the Process of Mixing

We must here remark upon certain peculiarities of the effect of mixing the various parts of the atmosphere. As we know it in the laboratory, mixing by mechanical stirring, which may be initiated or aided by thermal process, equalises the chemical composition and the temperature. That statement does not apply to the atmosphere. In the laboratory of the open air stirring certainly promotes uniformity of composition as regards the permanent gases; but it destroys the uniformity of composition in respect of water-vapour, and it also destroys the uniformity of temperature. Mechanical mixing tends towards equality of "potential temperature," i.e. the temperature as adjusted adiabatically to a standard pressure in accordance with the laws of dry air. The first result of a process of mixing is to warm the lower part and to cool the upper part. The cooling must result in cloud and rain if the process is carried on long enough. If it were not for the constant additions of heat which take place mostly at the surface and the additions of water-vapour which, apart from the evaporation of cloud, take place there exclusively, the perpetual churning of the atmosphere would result in complete uniformity of composition of practically dry air; the amount of water-vapour would be defined by the saturation pressure at the lowest temperature to be found in the whole mass, and there would be convective equilibrium up to the limit of height reached by the mixing forces. Every shake of a mechanical mixer would shake some water out of the atmosphere; but the artful stuff is always creeping back again.

4. THERMAL CONVEXION

Thermal convexion is really the process of sorting the atmosphere into layers according to their entropy. The entropy may be realised as potential temperature or latent as water-vapour. It will be convenient if we may call that part of the entropy which has been realised in the form of potential temperature "realised entropy." Incidentally convexion promotes the process of mixing, but its operation is somewhat complicated. Any specimen of air is directed and limited, as regards vertical motion, by the stratification of its environment as regards density, which depends upon pressure, temperature and humidity. It cannot penetrate a layer of which the lapse-rate of temperature is smaller than the adiabatic. If continuously warmed by the land- or the water-surface the first effect will be to create a layer in convective equilibrium; but, in time, that layer may extend so high and become so warm and so charged with moisture that it may pass the limit of equilibrium with its environment and travel through a succession of layers until a region of low lapse-rate is reached; and thus readjustment by convexion on a large scale supervenes. Any such readjustment results in the stratification of the atmosphere according to potential temperature not ordinary temperature, and the potential temperature may be increased automatically on the way up by the condensation of water-vapour. The potential temperature of a specimen of air is merely another form of expression of the entropy of its dry air, no allowance being made for the latent heat of its water-vapour. So long as the process of reduction to standard keeps within the application of the laws for dry air the formula $\log \theta_0/\theta = \dfrac{\gamma - 1}{\gamma} \log p_0/p$ holds, where θ_0 is the potential temperature for the standard pressure p_0, and the change of entropy from a standard temperature at pressure p_0 is $\phi - \phi_0 = c_p (\log \theta_0 - \log \theta)$. In the actual atmosphere the potential temperature may be increased by heat set free by the condensation of water-vapour, while the entropy of the mixture remains, strictly speaking, unchanged. Hence the use of the two terms is still desirable unless we may call the potential temperature the "dry air" entropy. Where the potential temperature or entropy of the environment increases with height convexion ceases to be possible.

5. THE COMPLICATION OF NATURE

In actual fact, instead of getting convective equilibrium and uniformity of composition we get continuous warming of air and evaporation of water at the surface, local convexion and irregular condensation of water in the upper levels followed by its removal as rain. The process is for ever going on; the atmosphere is for ever endeavouring by thermal convexion and eddy-motion to reduce itself to the state of convective equilibrium of dry air and for ever being balked by the constant supply of heat and evaporation of water at the surface.

6. The Simplicity of a Dry Atmosphere in Convective Equilibrium

If the atmosphere were dry the distribution of temperature would be simple. It would be regulated according to the adiabatic changes of dry air with a fall of temperature of ·97 a for each hundred metres. The mixing would extend as far as the thermal convexion of dry air could reach. That would depend upon the rapidity with which the upper air could get rid of heat, because any air warmed ever so slightly above its environment at the surface would begin to rise, and theoretically would find no resting-place short of the limit of the convective atmosphere. But in practice its charge of heat would be gradually dissipated by the turbulence of its relations with its environment on its journey, and although some capacity for upward motion would always remain, the rate of progress would be so slow that there would be in effect a stratification, with air in the upper regions somewhat warmer than the proper temperature for convective equilibrium. And thus the atmosphere below its tropopause would be in constant motion and in perpetual relation, as the working substance on the one hand and the working chamber on the other, of a vast heat-engine, of which the source of heat is a certain part of the surface and the sink of heat the troposphere and another part of the surface. The working substance would be that part of the atmosphere which received heat from the ground, the working chamber or environment would be the remainder of the atmosphere. The work would be represented by the kinetic energy of the air and any potential energy stored in the environment. Such energy exists where the air is potentially colder than its environment.

7. The Influence of Water-Vapour

This picture of the function of the troposphere is not impaired by the influence of water-vapour, but much accentuated thereby. The evaporation of water represents the conveyance of vast quantities of latent heat to dry air, which becomes active when condensation takes place and rain is formed in the upper layers. The heat so set free makes the actual reduction of temperature in rising air much smaller than that corresponding with the same reduction of pressure in dry air; and, as compared with an atmosphere of dry air in convective equilibrium, our atmosphere is, as a rule, much too warm at all levels above ground.

On this account the atmosphere becomes much more potent as a heat-engine. As we trace the temperature all the way up we find evidence of heat having been discarded at all levels, the temperature elevated in consequence, and the reversible cycle of dry air seriously impaired. These thermodynamic functions of the atmosphere are of essential practical importance. They regulate all the manifold operations of wind and rain. We shall not, therefore, have made a satisfactory representation of the atmosphere unless this vital characteristic of its function as a heat-engine is sufficiently brought out.

When we come to representations we are confronted by a difficulty. The supply of heat to the working substance and its subsequent elimination are subject to great local variations, some of them regular, such as the effect of latitude or the permanent distribution of land and water, some of them casual, such as the distribution of cloud or the particular condition of the surface.

Our only way of getting over the difficulty is to eliminate the casual conditions and generalise the phenomena, by dealing with mean values; and then, subsequently, to consider the characteristics of phenomena which are recognised as local in their operation and are therefore probably local in their causes. The generalised phenomena are known as the general circulation and the local phenomena include those of the travelling cyclone and anticyclone.

If this be the proper way of regarding the atmosphere we have to deal with the thermodynamics of moist air. That is simple enough, so long as the limit of saturation is not passed. It is passed so frequently in the atmosphere that the thermodynamics of saturated air must receive very ample illustration from the phenomena exhibited therein.

8. IRREVERSIBLE ADIABATIC LINES FOR SATURATED AIR

In dealing with the thermal changes in moist air under the influence of such changes of pressure as might be experienced on a passage through a vertical section of the atmosphere, upwards and downwards, we are faced with the difficulty that the sequence is reversible only if the condensed water is carried in the form of drops along with the air from which it was produced, a condition which cannot be satisfied when the elevation extends over several kilometres; the process ceases to be reversible, and is also somewhat modified, if the condensed water falls out. It certainly will fall out if the drops exceed a very small size: and, in fact, rainfall is the measure of the amount of water which owes its fall to the convective process.

Thermodynamic equations for dry air were among the earliest results of thermodynamic theory. An equation for the adiabatics of saturated air, assuming the condensed water to be retained, was given originally by Hertz[1] and in modified form by Neuhoff[2]. These are reversible adiabatics. The equation can be modified to allow for the water having fallen out and we then get a series of lines which have been called pseudo-adiabatics, but may be more usefully described as irreversible adiabatics. If the process of rarefaction on ascent be reversed by descent and the air gradually compressed the lines described on the return journey will be indistinguishable from the adiabatics of dry air. From our experience of rainfall we must conclude that in actual practice the condensed water is separated from the air which is rarefied by convexion and therefore that the irreversible adiabatics represent approximately the actual conditions for ascending air. During its passage upward the irreversible adiabatic of saturated air is described and on the

[1] H. Hertz, *Met. Zeitschr.* 1884, p. 421; *Smithsonian Misc. Coll.* vol. XXXIV, p. 198, 1893.

[2] O. Neuhoff, *K. Preuss. Met. Inst.* vol. I, no. 6, p. 271, 1900; *Smithsonian Misc. Coll.* vol. LI, p. 430, 1910.

downward passage the adiabatic of dry air modified perhaps by losses of heat on the way.

I present a picture (Fig. 48) of the reversible adiabatics of saturated air, with the effect of the freezing of water-drops omitted, which has been prepared for me by Mr A. W. Lee, of the Imperial College, and which is said to differ only slightly from the corresponding chart of irreversible adiabatics[1]. The finer lines in the diagram represent the reversible adiabatics. On the chart are entered the normal temperatures of air in the vertical at a number

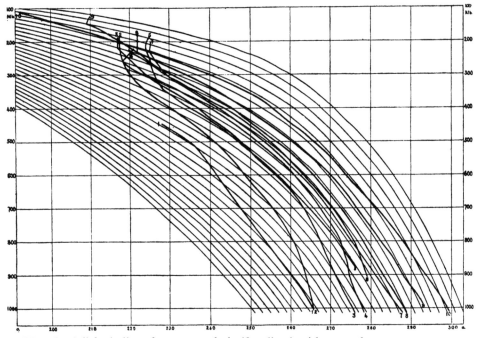

Fig. 48. Adiabatic lines for saturated air (fine lines) with normal temperatures at a number of stations (heavy lines). The stations are:

1. McMurdo Sound (summer). 2. Pavlovsk (winter). 3. England (winter). 4. Arctic Sea, 77° N. (summer). 5. Normal cyclone (England). 6. Normal anticyclone (England). 7. Pavlovsk (summer). 8. England (summer). 9. Canada (summer). 10. Batavia (winter).

of stations. These are shown by the heavier lines. They show that the relations of temperature to height in normal air and saturated air in favourable circumstances are very similar. The agreement between the lines of average temperature and the adiabatics of saturated air for the higher surface temperatures of saturation is very close until the stratosphere is reached. Such temperatures are found in tropical countries; but in higher latitudes the lines are not nearly parallel. Thus, not only do these lines indicate the changes through which every specimen of air must pass as it penetrates successive strata, but for the

[1] Mr Lee's estimates of the pressures on the reversible and irreversible adiabatics at temperature 223 a when initial temperature of saturation is 303 are 145 mb. and 140 mb. respectively.

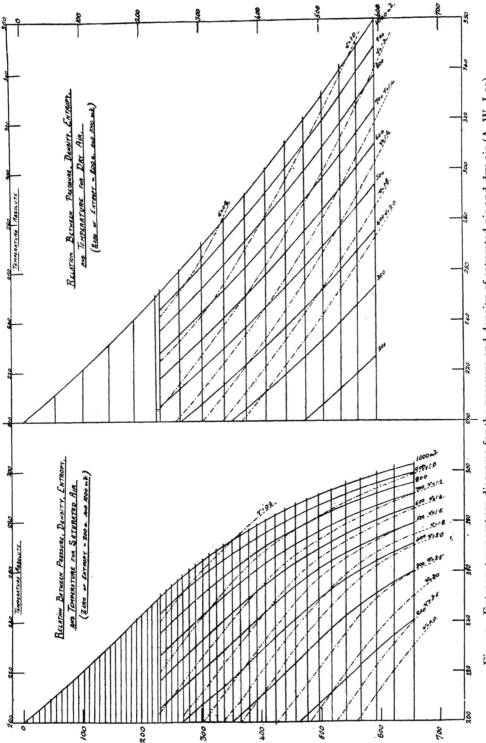

Fig. 49. Entropy-temperature diagrams for the pressure and density of saturated air and dry air (A. W. Lee).
The entropy is measured from 200 a and 1000 mb. as standard state, and is expressed in joules per kilogramme of dry air per tercentesimal degree.

higher temperatures with higher content of water-vapour they show the air in its normal state to be very little removed from the condition of instability, under which the path would be described in tropical latitudes; but in arctic latitudes the conditions indicate stability in a notable degree.

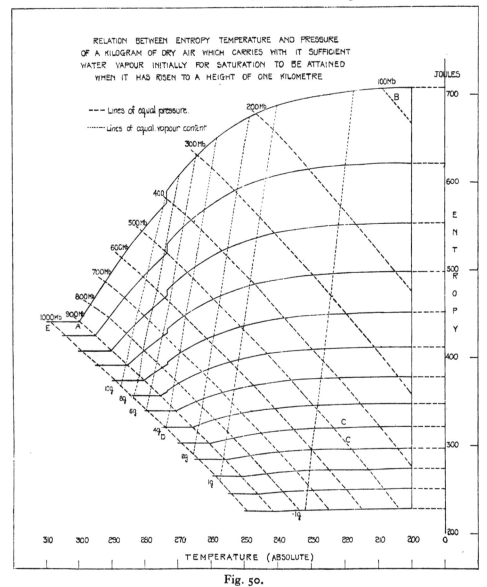

Fig. 50.

The diagram exhibits the relation of pressure and temperature under adiabatic conditions. In order that we may consider what happens in the atmosphere from the point of view of energy (and no other is ultimately of avail) we must change the independent variables and refer the properties either to axes of pressure and volume, which are usual in the text-books of thermodynamics, or to axes of temperature and entropy, which are more directly useful in the case of the atmosphere. For a complete representation of the properties of moist air we require at least two entropy-temperature diagrams, one for dry air and the other for saturated air. Diagrams of this character have been constructed for me by Mr Lee and are reproduced in Fig. 49. The values represented are explained in the legends of the respective diagrams. Here is an attempt to deal with the question approximately by a single diagram (Fig. 50), which I owe to Mr E. V. Newnham. It deals with the realised entropy of the air, that is, it shows the thermal changes in dry air when moisture carried with it is considered simply as a thermal agent, the loss of water being represented simply by the increase of realised entropy. It is sufficiently accurate to indicate the kind of framework into which the phenomena of the atmosphere must be fitted if we are to comprehend its winds as a dynamical expression of its thermal experience. We may infer from the study of this diagram that every kilogram of air which ascends to the stratosphere at the equator and returns to its old place on the earth carries 10 kg. units of heat to the stratosphere and on the way converts $2\frac{1}{2}$ kg. units of heat into work. Its efficiency is about $\frac{1}{4}$, and the efficiency of air rising in any other locality is less than that figure.

9. THE KINETIC ENERGY DEVELOPED

The output of the atmospheric heat-engine has thus been brought into the region of calculation and part of it is the kinetic energy of the winds which may be regarded as the atmosphere's fly-wheel. The amount is not easily computed but we may be certain that in modern scientific jargon it would be called "fairly large." I made an attempt to estimate the kinetic energy of the winds over the North Sea in the course of an examination of the weather that wrought havoc with our summer offensive in 1917. The mischief expressed itself as a small depression about 1400 km. wide and 11 millibars deep (Figs. 56–58). Its kinetic energy at its maximum stage worked out at $1\cdot5 \times 10^{24}$ ergs; that is sufficient to run 2360 million horsepower-days, or 6 million horsepower for a year. It is a mere fraction of the whole kinetic energy of the winds of the earth at any one time. Taking simple proportionality to area the kinetic energy of the atmosphere would be $2360 \times 332 \times 10^6$ horsepower-days, about 18 billion horsepower-hours. Truly it is a very big engine that carries that amount of energy in its fly-wheel. The daily work of this great engine is to make up for the natural degradation of the energy of the fly-wheel and incidentally to blow down a few trees, turn a few windmills, raise a few waves and help or hinder a considerable number of ships.

Here ends the general view of the structure of the atmosphere. You will notice that the working part of the atmosphere is the troposphere: the stratosphere comes in as a boundary or main deck, which confines the working substance at the top.

Let us proceed to consider some particulars.

10. AVAILABLE DATA OF THE ATMOSPHERIC ENGINE

I do not expect or intend this afternoon to give you a complete account of the method of working of the vast machine; but I will classify the information which I can give you under the headings of the boiler, the condenser, the working substance, which is the troposphere, and the fly-wheel or winds. I will interpolate certain considerations as to the natural resilience of successive layers of the atmosphere and the time-scale of different forms of convexion which appear to me to be essential to a proper understanding of the structure of the atmosphere. The high resilience of the stratosphere is suggestive of the idea of a piston under which the working substance acts.

11. THE BOILER AND CONDENSER

The boiler is that part of the surface of land and sea warmed above the temperature of the overlying air by the sun's radiation, and that part of the atmosphere, if any, which is warmed directly by radiation. The condenser is any part of the surface of land or sea colder than the air above it and any part of the atmosphere which is losing heat on balance by radiation. If we regard the troposphere as a black body radiating into space the energy of long waves and transmitting that of short waves we can look upon the troposphere as part of the condenser; but the phenomena of dew and hoar frost, and I should say also those of the polar front, warn us against regarding it as the whole. Boiler and condenser are indeed seen together in a very effective way on a map of the distribution of temperature over the globe (see Figs. 18–21, plates IV–VII). The characteristics of the boiler are most conspicuous over the surface of intertropical regions. The functions of the condenser are most clearly exhibited on high land exposed to the clear sky, as over Greenland, the Antarctic Continent and the slopes of mountain-chains that are in shadow. Heat is conveyed from the surface to the atmosphere partly by conduction of heat but still more by diffusion of evaporated water, and conversely heat is lost by conduction or by the formation of dew and hoar frost which may express a considerable part of the loss of heat over such a region as Greenland. Hence the distribution of moisture is to be regarded as of physical importance as well as the distribution of temperature. I exhibit, therefore, maps of the distribution of "dew-point" or "cloud-temperature" of the air over the globe for January and July (Figs. 51–54, plates VIII–XI). A very large area is included within the line of 295a which surrounds the regions favourable for convexion. There are a few isolated patches with dew-point 300a, where instability must be very probable.

12. THE WORKING SUBSTANCE OR TROPOSPHERE

The next view is a vertical section of the atmosphere from pole to pole (Fig. 55) exhibiting the distribution of temperature for midsummer in the northern hemisphere, and for midwinter in the southern hemisphere, by isothermal lines, the sections of isothermal surfaces which are nearly hori-

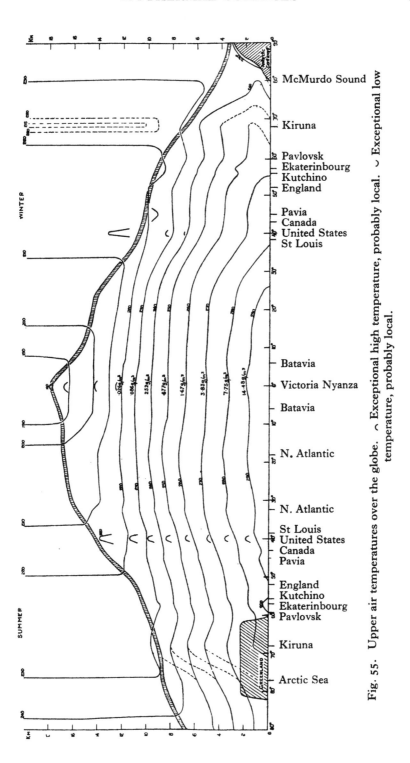

Fig. 55. Upper air temperatures over the globe. ⌢ Exceptional high temperature, probably local. ⌣ Exceptional low temperature, probably local.

zontal in the troposphere and nearly vertical in the stratosphere. It shows
the continuous line of section of the tropopause from pole to pole marked
by the sudden transition of the isothermal surfaces from horizontal to vertical,
the great range of temperature between the surface and the stratosphere at
the equator and the diminishing range and height as the equator is left and
the polar regions approached. The data in the vertical for different localities
are taken from all available publications in which summaries for localities are
represented. The gradual decrease of temperature with latitude which, from
our earliest schooldays, we have accepted as a general rule, is shown to extend
only to the level of 12 km. Beyond that level the higher latitudes have the
warmer air at corresponding levels; and, generally, within the stratosphere
the air is colder on the equatorial side of any vertical plane. Hence with the
highest temperatures of the surface at the equator we find the lowest temperatures immediately above our heads at the height of 17 km. Those in higher
latitudes have above them isothermal columns of higher temperature than
over the equator, and they are reached at lower levels, until at the summer
pole the genial warmth of about 240 a is possibly only about 7 km. overhead
instead of a frigid temperature of 190 a at 17 km. above the equator. At the
winter pole there is represented a stratification from the surface upward with
negative lapse-rate, in which may be recognised the working properties of
our *novum organum*, "the polar front." Above it the stratosphere descends
probably to its lowest limit, though the position assigned to it in the diagram
is conjectural.

The evidence in favour of the general rule of the dependence of the height
of the troposphere upon latitude is subject to two noticeable exceptions
among the observations: one comes from the observations of the Weather
Bureau of the United States, the other from the Franco-Scandinavian observations at Kiruna. I find no suggestion of a cause for the latter exception
unless the Atlantic coast-line is responsible. It is a very potent meteorological
agent and not far distant from Kiruna, but I do not know the details of its
operations. In explanation of the former I think we may possibly invoke the
general principle of the slope of the tropopause between high and low pressures
which has been worked out from the observations particularly by Mr W. H.
Dines, though it is now common ground. The tropopause is higher over high
pressure than over low pressure. In a cyclonic area, with the reduced level
of the tropopause is associated higher temperature than its environment in
the stratosphere and lower temperature in the troposphere; conversely, in an
anticyclonic area the tropopause is higher; the temperature is lower than that
of its environment in the stratosphere, but higher in the troposphere. In consequence there must be distortions of the tropopause and isothermal surfaces
associated with high and low pressures which can be best represented by the
model (Fig. 47). It shows that in the belts where cyclones and anticyclones
occur, certainly on the polar side and probably on both sides of the line of
tropical high pressures, a series of isothermal surfaces cut the tropopause and
turn up into the vertical above a local cyclonic depression forming a local

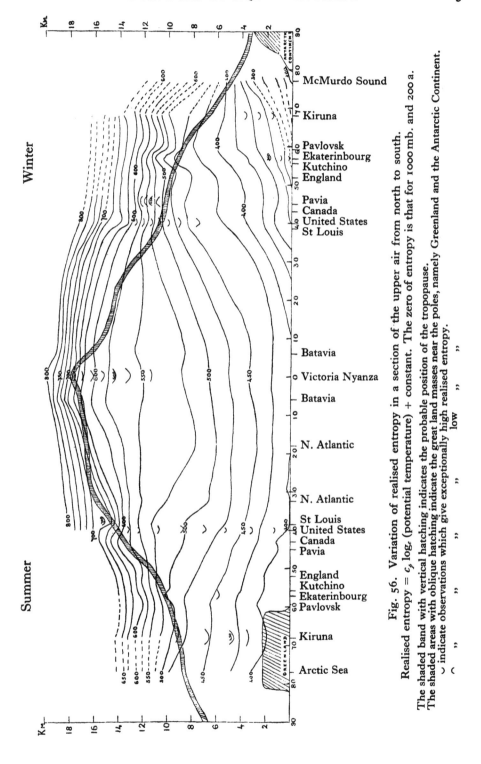

Fig. 56. Variation of realised entropy in a section of the upper air from north to south.

Realised entropy = $c_p \log_e$ (potential temperature) + constant. The zero of entropy is that for 1000 mb. and 200 a.

The shaded band with vertical hatching indicates the probable position of the tropopause.
The shaded areas with oblique hatching indicate the great land masses near the poles, namely Greenland and the Antarctic Continent.
) indicate observations which give exceptionally high realised entropy.
) (low „ „ „

"pole," while on the other hand over an anticyclone there is a local isolated pile of isothermal surfaces corresponding with what we see over the equatorial belt.

13. POTENTIAL TEMPERATURE OR REALISED ENTROPY

Let me now give another form to the analysis of the facts with regard to the distribution of temperature in the vertical. While in respect of temperature the stratosphere is apparently a cold region, in respect of potential temperature it is notably a warm region, or in respect of the realised entropy of dry air a region of maximum for its latitude. In the isothermal columns the potential temperature increases rapidly with height, approximately indeed, by ten degrees for every kilometre and the realised entropy of dry air increases to correspond with the increase of potential temperature (Fig. 56). Thus, in spite of its low actual temperature the uppermost region of the equator is still potentially much warmer than the surface, the air at the same level over the pole still more so.

With suitable reservations, the logarithm of the potential temperature is proportional to the realised entropy. And from a generalised point of view the operation of a thermodynamic engine is to get work out of a substance

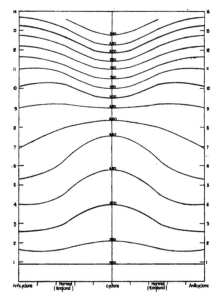

Fig. 57. Realised entropy in cyclone and anticyclone.

(The entropy is measured in joules from zero when $T = 200$ a, $p = 1000$ mb.)

by using it constantly to carry heat from the boiler to the condenser, to keep down the entropy of the boiler, enhance the entropy of the condenser and to transform into work the difference between the heat taken in and ejected.

Hence, the increased entropy of dry air in the upper strata is the by-product of the atmosphere's thermodynamic engine, and if we regard the dry air as the working substance we must regard the stratosphere as representing a marked enhancement of entropy due to heat originally derived chiefly from the ground, and the working of the engine as an incident in maintaining the stratosphere in a condition of high entropy. Hence the phenomena of weather must depend upon the details of the process by which dry air enhances the entropy of the upper strata. This statement is effectively illustrated by a diagram of the realised entropy of dry air according to Mr Dines's figures for high and low pressures (Fig. 57).

14. The Resilience of Successive Layers of the Atmosphere in
Consequence of Inversion of Temperature Lapse-Rate or any
Deviation of Lapse-Rate from the Adiabatic

As a necessary introduction to the question of the permanence of cyclonic circulations and the maintenance of their kinetic energy which forms a part of the atmospheric fly-wheel, I may here touch upon the question of what I call the resilience of the stratosphere, a property which the stratosphere enjoys in common with other isothermal regions, or regions of inversion of lapse-rate, in virtue of the departure of the lapse-rate of temperature from the adiabatic. Resilience is a general feature of the atmospheric structure which depends upon the deviation of the lapse-rate of the environment from the adiabatic lapse-rate of the rising air, and is therefore in particular a feature of the stratosphere.

An atmosphere in convective equilibrium is like a liquid of uniform density; it has no resilience at all; if any part is displaced to any extent it remains quite indifferent; but as soon as it acquires any "up-grade" of entropy it becomes, to a certain extent, a resilient medium with respect to the vertical motion of air. If air in vertical motion changes its temperature adiabatically, in a medium which has a lower lapse-rate than the adiabatic rate appropriate to the composition of the moving air, any vertical displacement of the air develops forces of restitution which are characteristic of resilience. With the very slight forces due to thermal changes in ordinary circumstances an isothermal layer, or still more, a layer with an up-grade of temperature, is highly resilient for air coming adiabatically from above or beneath. An ordinary spherical balloon may actually rebound from such a layer, as it would from an elastic surface, and so presumably likewise may free masses of warmed air that have appreciable vertical momentum. Similarly with regard to pressure. An area of low pressure formed in a region approximately isentropic can have its top or bottom "ceiled" by the resilience of a layer of air which has an up-grade of realised entropy.

In this respect a layer of air with high realised entropy at the top acts like a fluid of greater density below, and still more remotely like a solid, either above or below. The heavy fluid owes its resilience to the effect of gravity, a solid ceiling, to rigidity opposing gravity, the solid ground to rigidity reinforcing gravity. Thus we may regard the stratosphere as forming a resilient boundary for columns of low pressure, just as water, and still more effectively the solid earth, forms a lower boundary for such columns. Dynamically the difference is merely one of scale. A column, with a core of low pressure 10 mb. below its environment, passing over the sea would get its lower "ceiling" by the lift of the surface of the water through about 10 cm.; when passing over land, the deformation of the solid surface by one-hundredth of a millimetre would, I suppose, suffice. Passing under the stratosphere, according to Mr Dines's figures, it would pull the stratosphere down by about

1 kilometre in order to secure the necessary ceiling; clearly there is nothing outrageous in the suggestion, for with the figures which I have given the ratio of the resilience of water to that of the land is represented by $1:10^4$ and the resilience of the stratosphere to that of water is represented by precisely the same ratio. Whatever minifying glass you use to bring water into relation with land will equally bring the stratosphere into relation with water. Once more it is only a question of units; if your unit of resilience is small enough, the tropopause is equivalent to solid earth in its behaviour towards travelling columns of low pressure. In like manner the normal stratification according to potential temperature or realised entropy with the higher values on top gives a certain amount of resilience to the atmosphere and a corresponding tendency towards stratification in other things such as clouds and air-currents. It follows therefrom that air forced upwards by convexion makes a local break in the continuity of the layer as though it were breaking through a ceiling. It does not cause such wide disturbance of the general structure as would be caused if the law of convective equilibrium were operative.

15. The Time-Scale of Vertical Movements

If we regard the atmospheric processes as dependent upon the maintenance of entropy at the base of the stratosphere by means of heat derived by convexion, as distinguished from molecular diffusion, conduction or radiation, it is obviously important to consider the rate at which convexion can take place, and this is limited by the rate at which air rises or falls. When the anticyclone acquired its name it was regarded, and has been regarded ever since, as the reciprocal of the cyclone, with clockwise rotation in place of counter-clockwise, high-pressure centre instead of low-pressure centre, outward flow of air in place of inward flow, and, as a natural consequence, descent of air in the interior in place of the ascent of air. As cyclones and anticyclones undoubtedly form a characteristic feature of the structure of the troposphere in middle latitudes, the question of the ascent of air in cyclones and its descent in anticyclones belongs to the general question of atmospheric structure.

In this paper, as in others, I have assumed that in the upper air the motion of the air balances the distribution of pressure except in so far as some convergence towards the pole is necessary, on the one hand, to make up for the loss of kinetic energy in the body of the fly-wheel through air-friction, and, on the other hand, by the law of continuity of mass, to make up for the drainage of air down the cold slopes of the polar regions. Also, be it said, the control of motion by pressure becomes feeble in the equatorial regions and vanishes at the equator altogether. My reasons for accepting this position are set out at length in *Manual of Meteorology*, Part IV, and may here be briefly summarised by the observation that the tendency of wind to set along isobars must be extremely potent in the atmosphere because it disclosed itself to observation at the surface where conditions are extremely unfavourable for it, and the unfavourable conditions do not extend to the upper air. I assume, therefore, that to determine the inclination of the wind to the isobars in the upper air of latitudes beyond 40° we need only consider the convergence towards the pole. Judging from the spread northward of the Krakatoa dust, the movement poleward is real, but involves a journey of months rather than days. On the other hand, the circuit round the earth would be complete in, say, 15 days; hence the angle of the imaginary spiral round the earth marking the progress of air to the pole would be very nearly a right angle. In other words the wind would agree very closely with the isobars as regards direction.

The acceptance of this principle enables us to compute, from the evidence which weather-maps and other data afford, the rate of rise and fall of air in regions respectively cyclonic and anticyclonic. Judging by the size of hail-stones which are presumably carried up by an ascending current the rate of upward motion must on occasions be extremely great, the terminal velocity of a hailstone 1 cm. in radius is 22·5 m/s. But in the ordinary way an upward velocity for the air which carries a pilot balloon in the lower strata only reaches or exceeds the velocity of the balloon itself on 25 occasions out of 1000, and descending air moves as a rule more slowly than ascending air.

It can easily be shown that in a region where the law of velocity is that of the simple vortex, viz. $vr =$ constant, or in a region of straight isobars, if there be a flow across isobars from high to low, as nearly always there is, the surface-stream will pass under the upper current without any upward convexion if the angle of incurvature be everywhere uniform. And in like manner, it can be shown that in any region where velocity is proportional to distance from a centre, $v/r =$ constant, as in the form of a cyclone to which I once gave the unsatisfactory name of "normal," or approximately in any anticyclone, a flow across isobars with uniform angle of incurvature requires a uniform elevation or settlement of the air over the whole area included within a circular isobar. The amount of the elevation or settlement can be calculated from the linear dimensions, the winds, and the angle of incurvature if we are permitted to assume that the outflow is a phenomenon belonging to the surface which diminishes as the surface is left behind and is zero at one half-kilometre.

On this, or similar, supposition I have roughly computed the outflow from the "hyperbars" or great anticyclone of the Atlantic, and find it gives a settlement to the extent of 86 m. per day. Such a settlement would mean slow descent which, starting from 10 kilometres, would require 120 days for completion and would therefore indicate conditions of stability much more nearly in agreement with the general character of an anticyclonic area than the rapid local descent of air, which is connoted by "downward convexion," and has generally been understood as occurring in an anticyclone. An example of a local anticyclone over the British Isles on 26 March, 1907, gave a rate of descent of 264 metres per day.

On the other hand I have computed the flow into the depression of 3 August, 1917, already referred to and it indicates a general elevation over the area at the rate of 150 metres per day; which is of the same order as the anticyclonic descent. Such a rate of inflow, I compute, would fill up the depression in $2\frac{1}{2}$ days, and, according to the maps, that particular depression actually filled up on the spot in 3 days (Figs. 58–60). This is a matter of some importance in the theory of cyclones. If we take, for example, the figures for temperature in high and low pressures given by Mr W. H. Dines, *Character-istics of the Free Atmosphere*[1], p. 69, we find that air saturated at 280a would not penetrate the layers of air in low pressure and therefore ordinary convexion would require either saturation at a higher temperature or an environment more nearly isentropic; but inflow would cause some elevation and sufficient perhaps to account for a substantial part of the margin of differ-ence between the excess of temperature of high over low at corresponding levels and the dynamical change corresponding to pressure-difference. The correlative "dejection" in high-pressure areas would account for the rest of the difference.

Again, on the basis of the rate of ascent of pilot balloons and the law of resistance according to the square of the velocity, and making no allowance for turbulence due to the relative motion, I have computed the rate of ascent of warm air to be 12 metres per minute for 1a of difference between ascending air and environment, 17 metres per minute is permissible for 2a, and with those velocities, air would accomplish the journey across 10 kilometres of height in about 14 hours or 10 hours, and so on respectively.

If these tentative suggestions are found to hold we must distinguish between the local convexion that penetrates layers of air and causes heavy rainfall in a cyclone or elsewhere, or conversely, a cataract of air in certain localities, and the general elevation in a cyclone or settlement in an anticyclone which are the results of inflow at the bottom or outflow there, as the case may be, although these may come under the general name of convexion, and, so far as the cyclone is concerned, be the originating causes of the more vigorous examples of convexion which are characteristic of such regions.

So far as the descent of air in an anticyclone is concerned the motion can only fairly be described as "descending air" if the month be the unit of time; and in a cyclone the case is only different in so far as the inward motion would fill up the cyclone in the course of hours unless some additional energy were provided, and for the additional energy we depend upon real convexion.

I should like to distinguish between these two forms of the upward and downward motion of air and to limit the use of the term "convexion" to the cases in which the rising or falling air penetrates into, and passes through, the layers of air above it or below it, and use the terms "elevation" and "settlement," or perhaps "dejection" as a better correlative of elevation, for those cases in which air accumulates in gradually increasing thickness or the reverse takes place without any definite penetration of the layers above or below. I do so for two reasons. The first is that the proposed names enable us to describe the phenomena of the stratosphere with greater clearness. The stratosphere is not subject to convexion, but is subject to elevation or dejec-tion by local accumulation of air or by the removal of air from below. The

[1] *Geophysical Memoir*, No. 13. M.O. Publication 220 c, 1919.

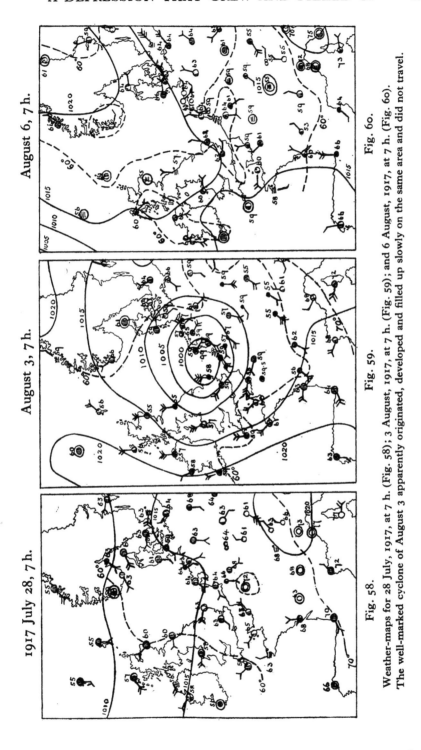

Fig. 60.

Fig. 59.

Fig. 58.

Weather-maps for 28 July, 1917, at 7 h. (Fig. 58); 3 August, 1917, at 7 h. (Fig. 59); and 6 August, 1917, at 7 h. (Fig. 60). The well-marked cyclone of August 3 apparently originated, developed and filled up slowly on the same area and did not travel.

second reason is that, if I understand the matter correctly, convexion, whether upward or downward, necessitates a certain amount of "eviction" of air from the layers through which the convected air passes; but elevation and dejection have no such accompaniment. I have dealt elsewhere with the effect of eviction in producing cyclonic circulation and have recently illustrated the process by a working model (see pp. 103 ff.).

16. THE FLY-WHEEL

We can now turn our attention to the distribution of winds which form the fly-wheel of the atmosphere. For this purpose we are unable to utilise to any large extent the direct result of observations of the winds in the upper air which have now become very numerous, because any single observation is a composite effect produced by local influences causing perturbations of the general or normal conditions. The analysis in the individual cases is impracticable; but for the atmosphere as a whole we can deal with the normal conditions by taking the average distribution of pressure at successive levels and computing therefrom the winds which correspond. The distribution of pressure can be computed from the temperature conditions by subtracting the pressure of successive layers from the pressure at the surface for which the distribution is well known, or by adding on pressure for successive layers from the level of 20 kilometres downward for which the pressure, as noted by Mr Dines, is, so far as we can tell, very nearly uniform and equal to 57·5 mb. A series of maps for successive steps of 2000 metres from the ground up to 20 km. have been computed for me by Miss Austin; for the lower 10 km., and for the equatorial "half" of two steps further, by subtracting from the surface pressure; and for the upper 6 km. of the polar half by adding computed pressure to the assumed uniformity of the level of 20 km. The map for the 8 km. level is shown in Fig. 61 (plate XVIII), while Fig. 62 (plate XIX) gives the pressure due to the

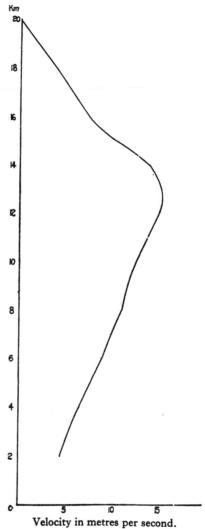

Velocity in metres per second.

Fig. 63. Variation of wind with height over England (calculation from the distribution of pressure).

lowest 8 km. computed as the difference between the pressure at the surface and the pressure at 8 km.

For each of these ten maps the normal velocity of the wind can be calculated and we get a view of the average kinematics of the atmosphere (Fig. 63). We notice that the fly-wheel regarded from this point of view consists of three parts. (1) A general westward circulation with some breaks in the equatorial regions between N. and S.; (2) a general eastward circulation from 30° to the pole which degenerates into (3) a complicated circulation in the surface-layers which we know to be stable in some latitudes, unstable in others and with which the kinetic energy of cyclones and anticyclones may be associated. The variation of the velocity of the wind with height from observations made at Batavia is shown in Fig. 64; the observations show a remarkable increase in the velocity from the surface to 14 km.

If the details of the maps are to be trusted the intertropical part of the fly-wheel should not be regarded as a continuous horizontal belt with increasing velocity upward, but as a number of local circulations having a connecting easterly stream at the equator on the south, which brings the two hemispheres into communication, and a connecting westerly stream on the north which establishes communication with the circumpolar circulation. Between these two main streams in certain localities are north and south streams established apparently by the influence of coast-lines, above which are bridges of communication between the easterly and westerly drifts. The polar part of the fly-wheel is a great horizontal circulation round each pole with a maximum of velocity in the upper middle layers and two comparatively calm regions at top and bottom. The bottom region is, however, disturbed by local systems to which we must give attention as they represent the third part of the fly-wheel dependent upon the variable cyclonic circulations. Parenthetically we may note an interesting question, whether the main equatorial and polar parts of the energy of atmospheric motion are really independent. If we regard the motion as relative to the earth's surface it takes as much energy to build up a current of east wind as of west wind; but regarded as relative to axes "fixed in space," if such a phrase may be allowed, east wind becomes a diminution of the motion of an atmosphere rotating with the earth and west wind an acceleration of that motion. From that point of view the energy of the polar circulation may be the equivalent of so much energy withdrawn from the equatorial circulation. Perhaps in the course of the discussion relativists may tell us whether there is an organic connection between the energies of these two circulations.

It should be noted that reasoning from the variation of pressure with height we have been led to the conclusion that at 20 km. there is practical uniformity of pressure and consequently no geostrophic wind. This conclusion does not prevent the extension of the east-to-west circulation over the equatorial regions to the level of 20 km. and beyond, because the motion of the air of the equatorial region is not subject to geostrophic control.

With what happens beyond the level of 20 km. we are not immediately concerned to-day; but we may remember that the process of computing pressures at successive levels from the distribution of temperature does not end at 20 km.

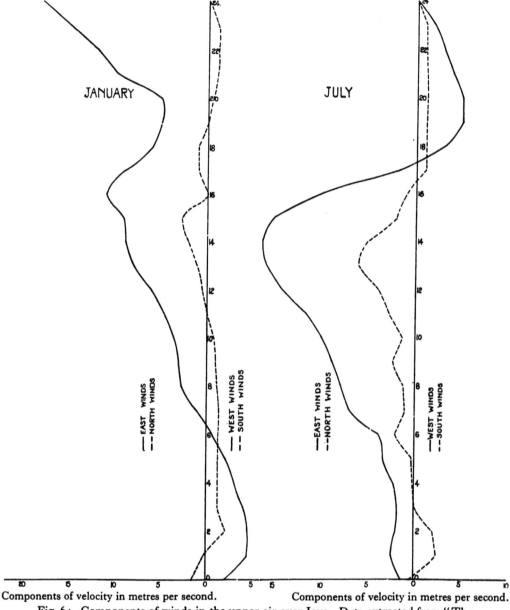

Components of velocity in metres per second.   Components of velocity in metres per second.

Fig. 64. Components of winds in the upper air over Java. Data extracted from "The atmospherical circulation above Australasia according to the pilot balloon observations made at Batavia." (Van Bemmelen.)

and if pursued beyond that limit the relations of high and low pressures to the equator and poles would be reversed. If, therefore, the isothermal surfaces remain vertical the fly-wheel of the equatorial belt will extend in the upper regions on either side to cover the whole globe. In that case the pressure-difference will increase with height and the corresponding velocity of rotation will gradually increase. Hence the earth may be rotating within an outer shell that has no share in the rotation. That, however, carries speculation beyond our present subject. We have very few indications of velocity in those high regions; the observations of Krakatoa dust are the best known. Apparently the velocity at 80 km. was about 80 statute miles per hour[1]; the dust took $12\frac{1}{2}$ days to get round the earth, at a height varying from 40 km. to 20 km., a long way short of round the earth in 24 hours.

17. The Stratosphere

So far the information which we have adduced about the stratosphere up to the level of 20 km. leads to the conclusion that it adds its quota to the two general circulations round the polar axis which go on beneath it and which constitute two of the three fly-wheels or parts of the generalised fly-wheel. Its property of practically isothermal structure in the vertical agrees very well with a theoretical curve obtained by Emden for the distribution of temperature in a quiescent atmosphere necessary to preserve the balance of thermal radiation. We have seen that the stratosphere cannot be regarded as entirely quiescent but its motions are stratified in such a way as not to interfere with the thermal distribution. There is no sudden discontinuity of motion when the tropopause is crossed upwards or downwards. Its chief function is to act as a "deck" for the troposphere beneath it. According to the observations, however, it is subject to temporary and local changes of temperature as changes of pressure pass underneath it, or for some other reason. The range of temperature is of the order of 20 a. We may therefore describe it as free from convexion of the penetrative type but still subject to air-displacements of the cumulative type producing either elevation or dejection.

So far as its motions are concerned we conclude from the distribution of pressure that the average motion is circulation round the pole as indicated by the maps of pressure, and rotation with the equatorial belt. We have, how-ever, a certain number of observations of wind-velocity in the stratosphere, some of which are figured in *Manual of Meteorology*, Part IV, p. 76, and others in a paper by Major G. M. B. Dobson in the *Journal of the Royal Meteorological Society*, vol. XLVI, from which Fig. 65 is taken.

There appears to be general agreement that wind falls off rapidly in the stratosphere down to a limit of small velocity and in so far as the winds are cyclostrophic the conclusion is in accordance with theory[2]. The limiting velocity is, however, not reached until a level of 13 to 15 km. is attained, 2 or 3 km. beyond the normal level of the tropopause. Another characteristic brought out by Mr Dobson's figures is that winds from north-east are as

---

[1] 70·4 m/h for the first circuit, 76·4 for the second, 76·3 for the first half of the third. *Krakatoa Volume*, p. 428.

[2] *Manual of Meteorology*, p. 90.

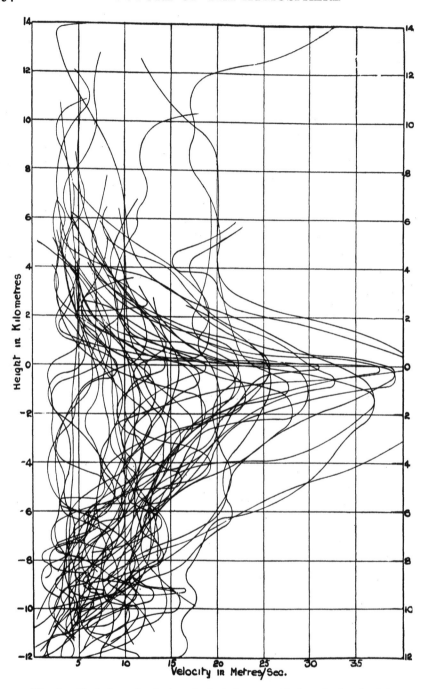

Fig. 65. Variation of velocity of wind above and below the level of the tropopause (Dobson).

The datum level for height marked o is the level of the tropopause for each of the several soundings.

frequent as winds from south-west. Perhaps, therefore, we are justified in regarding the limiting value of the velocity as the residual rotation of the stratosphere and the stronger winds below as parts of cyclonic or anticyclonic circulations extending beyond the normal level of the tropopause and affecting the air of the stratosphere itself. Such conditions may be regarded as an inevitable consequence of the process of ceiling a cyclone or anticyclone and perhaps we may consider the stratosphere over a cyclone as drawn down

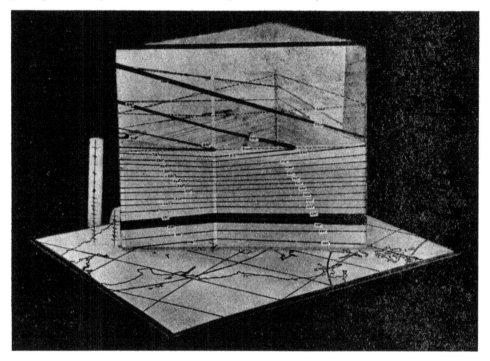

Fig. 66. Temperatures and pressures in a block of atmosphere 15 miles thick over a triangular portion of the British Isles.

From observations taken on 29 July, 1908. Block seen from the north-east. Isotherms are shown for each 5 a (tercentesimal) from 285 a to 205 a. The space between the isotherms of 270 a and 273 a is filled in; for other isotherms a thickness corresponding with ½ a is covered. The beaded lines in the stratosphere are isobars for 200 mb. and 100 mb. respectively. The arrows on the standards face the wind as determined by observations with theodolites.

through a kilometre or two kilometres in order to ceil the cyclone and as affected for two kilometres higher up by the local reduction of pressure and consequent cyclonic circulation. On the other hand the accumulation of convected air which has reached its limit of ascent must entail an increase of pressure in the troposphere and an elevation of the stratosphere.

We are faced with a difficulty arising from a deduction to be drawn from the assumption of strict verticality of isothermal surfaces. Just as we have argued that, with vertical isothermal surfaces, general circulation from east must set in over the layer of calms which is found at the level of 20 km., and increase with height, so we may argue on the same ground a corresponding reversal over high pressure and low pressure if these areas terminate below the general layer of rest at 20 km. To avoid that difficulty we want to assume that in the cap of an anticyclone within the stratosphere the temperature gradually increases and in the cap of a cyclone it gradually decreases in order that the differences of temperature, pressure and velocity may all fade out together and become continuous with the undisturbed medium with its normal rotation. This would mean greater complexity of structure in the stratosphere as regards temperature than is usually assumed. It may possibly be real, for some complexity in the right direction is shown in the models, *Proc. R. I.*, 1916, pp. 20–1; one of these is shown in Fig. 23 and the other in Fig. 66[1]. The occasions for which observations of the stratosphere in adjacent localities exist in sufficient number to decide this question and many others which have been indicated in this sketch, are very few and will not increase unless we see to it. The obvious conclusion to be drawn is that our maps of the world need further observations with ballons-sondes, particularly from great areas of ocean, desert and perpetual snow. Some have already been made at sea, the technique is well tried. With aeroplanes as scouts instead of ships, it might now be extended to tropical deserts and perpetual snows.

*   *   *

Note on a measure of the prevalence of convexion in the atmosphere.

From a discussion of the lapse-rates in various European countries it has been inferred that the flow of heat in the air of middle latitudes is downwards, and thermal convexion is not operative in the conveyance of heat from the ground upwards (W. Schmidt, *Met. Zeitschr.* Sept. 1921). That these statements should be true as regards average conditions is not surprising. For if we consider rain, which is not merely the drizzle of prolonged turbulence, as an indication of convexion, experience shows that there are relatively few occasions when it is operative. According to a table in *British Rainfall*, 1920, p. 35, the greatest total duration of rainfall in any year from 1881–1915 inclusive is 688·5 hours, or two hours in 25. The average is only one hour in 20, and in 1898 it was only one hour in 29. Considering that these hours include those of cumulative convexion (p. 128) and the drizzle of prolonged turbulence (*Manual of Meteorology*, Part IV, p. 50) very few hours are left for the study of penetrative convexion, perhaps one in 50.

---

[1] There is remarkable agreement between Figs. 66 and 47 in their representation of an anticyclone as surmounted by an intrusion of cold air from the south and with an inversion above it.

The two models were constructed quite independently and the agreement is probably more than mere coincidence.

9. THE AIR AND ITS WAYS

REDE LECTURE, 9 JUNE, 1921

A PUBLIC discourse on the atmosphere is a hazardous undertaking. On the one side there is the rock of obnoxious technicality and on the other side a whirlpool of iteration of the commonplace for which the classical adjective is damnable. The subject is very wide; it is, indeed, all-embracing. Everybody knows something about it, even the youngest. Once on my way home from a day's work on the commonplace problem I found a small tot, aged about six, playing in the road with a still smaller tot about three, and as I passed them dark clouds were gathering. The elder broke off the game with "Come along quick, Gladys, it is going to rain." And it was so. If it is possible to get so far at six how far ought one to get by sixty? Many of us, doubtless, have some weather-wise friend or acquaintance who really knows all about it and in plain language too. Quite the most perplexing people I know are those who aver that they have an exceptionally good barometer that succeeds when other people's barometers fail. Freed from obnoxious technicalities, if we may trust Mr Punch, the subject is summed up in the question "Will it rain to-morrow?" Every day since 1 April, 1879, an official answer has been given to that question. Yet it remains the most intriguing question about the air; but undeniably commonplace. Let us avoid it for to-day.

1. THE PECULIAR WAYS OF THE FREE AIR

On the other hand there is the risk of obnoxious technicalities. The lack of science in education partly accounts for the obnoxiousness, but not entirely; something must be said about the aloofness of science itself from the ways of the air. Before the establishment of the principle of conservation of energy in the middle of the nineteenth century the natural philosophers of this and other Universities used to find in the phenomena of the atmosphere their chief illustrations of the principles of experimental physics. Senior members of the University will remember the quaint survivals of the practice in the papers on *Heat* in the old General Examination and the mild efforts which we made to enable our pupils to satisfy the examiners. Chronologically the development of the daily map of the weather, the working chart of modern meteorology, and the establishment of the principle of conservation of energy came together. If the old tradition had held, the weather of the map would have furnished examples of the application of the new principle; but for the most part the natural philosophers have passed by on the other side and have looked elsewhere, with conspicuous success, of course, for illustrations of physical laws. Meanwhile the physical problems of the weather-map have not been solved.

The subject is inherently difficult. In the first place, the atmosphere is on such an immense scale that its behaviour is not to be brought under the principles of physics without much trouble, and, I may add, many mistakes. The most confident theories of the past are flatly contradicted by facts which have come to light since the investigation of the atmosphere was extended to the upper air by balloons, kites, kite balloons, and now by airships and aeroplanes. We have now many facts at our disposal about the atmosphere up to 20 kilometres. They are, of course, not necessary for the formation of a correct theory, because no new principles are involved; but they are invaluable for the purpose of verification or contradiction by which hypotheses get moulded into consistent theory. The behaviour of air in bulk is so entirely different from that of the laboratory sample that the ways of the air are, indeed, as peculiar as those of "the Heathen Chinee." The air as we know it in the laboratory is a very mobile fluid, yet in the atmosphere it manages to take on a sufficiency of the character of an elastic solid. It does not go in the way it is pushed. Pushed north it goes east and pushed east it goes south. The condition for its making a journey to the north is that it should be pushed west. If you blow a jet of air straight upward you may find that part of the effect is a vortex whirling around you. In front of its fire, the sun, the air will very likely get colder instead of warmer as you and I would get; losing heat by exposure to the clear sky on a cold night it may get warmer. In spite of all that is taught in the laboratory about the levitating effect of warmth, cold air floats above us with warmer air beneath. If you tell the air that warm air rises it winks an eye and interjects an "if" and a "when." If the Olympian gods felt cold and thought to make themselves warmer by stirring up their chilly air with the warmer air enjoyed by mortals down below them they would be disappointed. Stirring would make them colder and us warmer. Shake air up violently, water falls out of it; and if the shaking went on long enough the air would become intolerably dry, very cold at top, very warm at bottom. Not only has the air the innate capacity for these conjuring tricks but it never, or hardly ever, fails to use them. I was not aware of these peculiarities of the air when I left Cambridge for the airy occupation of an office in London; but I know them now.

2. THE GENERAL PROBLEM OF THE SCIENCE OF METEOROLOGY

Yet, underlying the work that is done in meteorology officially or unofficially there is, and has been all the time, a definite purpose to bring our knowledge of the air into relation with the laws of physics as established in the laboratory and therefore particularly with the laws of energy. I shall endeavour to justify this statement by what I have to say, steering, if I can, a course between the Scylla of technicalities and the Charybdis of commonplace. I shall ask you to believe with me that that is the way to find the ultimate answer to the question "Will it rain to-morrow?" Let us regard the question "Why did it rain on Tuesday?" as one which in logical order comes first, and must, in fact, be answered before the future can be predicted with certainty. If you agree, we shall have to begin at the beginning once more and I shall lay before you what I fear you will consider most elementary facts; but many of them at least are in a new dress. I am afraid I cannot hope to be quite coherent in what I have to say; but let me make it quite clear to begin with that there are two sides to the study of the air and its ways, which can be pursued by different people who may never meet each other. One is the observation and collection of the facts about the weather for every part of the world, the other is the interpretation of the facts by dynamical and physical reasoning. Nothing, at least nothing useful, can be done without real facts; but real facts do not, as a rule, explain themselves. The subject is, once more, peculiar among scientific subjects in that every worker upon theory must rely upon observation by others for his facts. That in itself is rather deterrent; but not even the most active observer can supply the needs of his own theories. Every one of us is, therefore, an out-and-out free trader in information about the weather. The collection of facts is the result of international co-operation beginning from the seventeenth century, when Halley published his map of the winds of the globe and Dampier described the phenomena of a typhoon. It started afresh in the nineteenth century when Alexander von Humboldt put together observations of temperature to make a map of the distribution of that important element over the globe; it is now the subject of an international organisation as to methods of observation and the presentation of results, which is common to the whole world.

3. THE FUNDAMENTAL FACTS

I shall show some of the results, arranged, according to custom, in such a way as to enable us to draw some general conclusions from many millions of separate observations collected over a long period of years. Normals we call them, though it is perhaps not easy to justify the word.

The composition of air

As a preliminary and at the same time an example of the need for caution about fundamental facts let me show what the air is made of by pointing to three results for the composition of the air at successive levels up to 400 kilometres according to Washington, Marburg and Cambridge respectively (Fig. 67). According to the first the top of the atmosphere is hydrogen, to the second geocoronium, and to the third helium. The differences are obvious; they depend to some extent on refinements of spectroscopic observation, but mainly upon what inferences one can draw from the fact that the estimate of free hydrogen to be found in the air may vary from ·02 to ·003 per cent. by volume.

Below the level of 20 kilometres we are not troubled with questions of composition. Everything is well mixed except in regard to the amount of water-vapour, a very important exception, since as I have already hinted when you churn up the air water falls out.

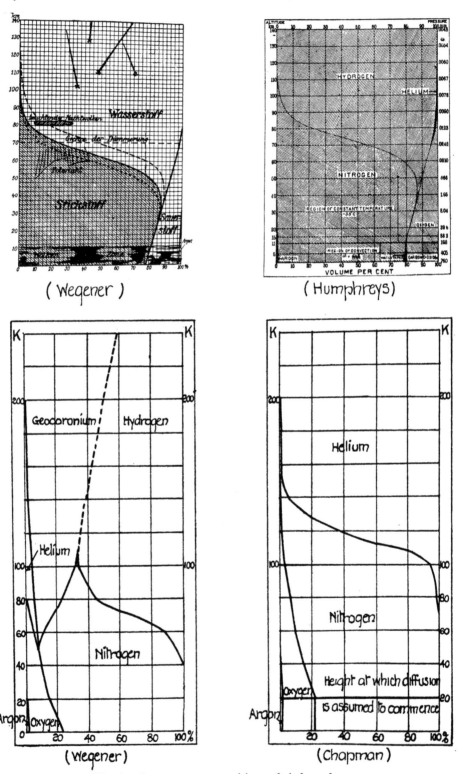

Fig. 67. Percentage compositions of air by volume.

The distribution of land and water

My first illustration is of the geographical features; my map shows the coast-line dividing land from sea and also the contours of 200 and 2000 metres (see Figs. 14 and 15). Those contours would become coast-lines in case of deluges of appropriate dimensions. The line of 200 metres is in a way the effective coast-line of the air. I ask you to bear in mind particularly Greenland, the Antarctic Continent, the line of the Rocky Mountains and the Andes, and the roof of the world in Central Asia. We may have occasion to bear them in mind as geography's contribution to the complexities of the atmospheric problem.

The normal distribution of temperature at the surface in summer and winter

My next maps represent the distribution of temperature of the air in summer (Figs. 19 and 21, plates v and vii) and in winter (Figs. 18 and 20, plates iv and vi). They summarise the main aspects of the conditions produced by solar and terrestrial radiation. There are some subsidiary aspects which are also of importance and express themselves in the temperature of the upper air which I shall refer to later. These maps differ from the commonplace in leaving the lines of temperature of the air over sea and over land respectively, quite untouched by any attempt to join them up and so conceal the discontinuity of temperature which is characteristic of coast-lines and is in some places very conspicuous. That omission is deliberate. It is intended to emphasise the reality of coast-line as an agency in the control of the air and its ways.

The normal distribution of water-vapour

My next views are of the distribution of water-vapour as represented by the dew-point of the air or the temperature at which cloud would form (Figs. 51–54, plates viii–xi). It is easily convertible into the quantity of water in a cubic metre. High dew-point means much water in the air. Water-vapour is the chief agency of nature for conveying heat to the atmosphere and this map therefore represents the driving force of the atmosphere in its chief aspect. Upon it are dependent the possibilities of cloud and rain and all that those small words connote in relation to weather. For the sake of brevity I am confining your attention to the summer month, July, and shall do so henceforward. Note the extent of the area within the line of 295 a on the centigrade scale measured from 273 a below the freezing-point of water. That line encloses the region where water-vapour is most abundant. It represents a focus of thermal activity. It is accentuated in a few localities by lines of saturation at 300 a (80·6° F.)—very moist and therefore very intemperate regions.

Note in contrast how very little water the air of the arctic regions has to dispose of. It cannot be from those regions that we draw our supplies of rainfall, yet they take their part in the machinery which produces it.

The contrast between the equatorial and polar regions in this respect is very clearly manifest in a diagram (Fig. 68) which indicates by the number of

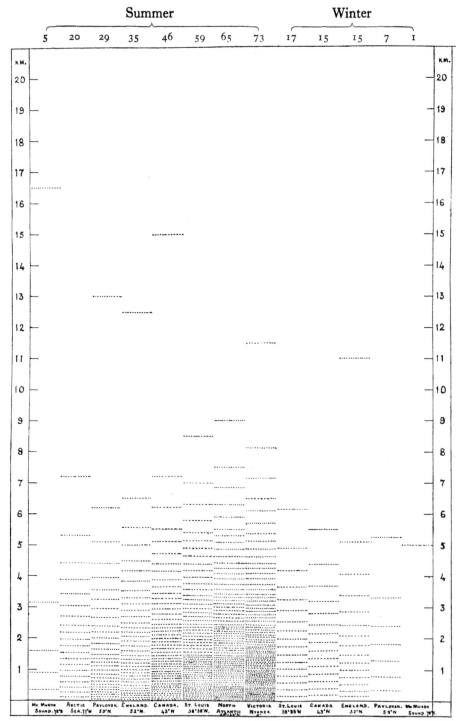

Fig. 68. Maximum water-vapour content of the atmosphere at selected stations (calculated from observed upper-air temperatures, assuming saturation).

The number of lines in a vertical column gives the number of kilogrammes of water-vapour in a column of cross-section 1 sq. metre, or the amount of rain (in millimetres) which would fall if the atmosphere were completely desiccated. The drawing of the lines is based on temperatures corrected to sea-level.

dotted lines in vertical columns the amount of rainfall in millimetres which could be contained in the atmosphere over the respective stations named in the diagram. The numbers are entered at the tops of the columns.

The normal distribution of cloud

Next, let us look at the average distribution of cloud in the Northern and Southern hemispheres in the same month (Figs. 69 and 70, plates XII and XIII). It is indicated by lines which connect together the points of the earth according to the average proportion of sky, estimated in tenths, which is covered by cloud during the day. There are nearly cloudless regions, less than one-third of the sky covered, and very cloudy regions more than seven-tenths of the sky covered.

The normal distribution of rainfall

Next, we proceed to a map of the distribution of rainfall in the Northern hemisphere in July (Fig. 71, plate I). This may shock you very properly as a ludicrously inadequate map because the information refers exclusively to the land-areas; it tells us only little of the sea; we have unfortunately so few observations of rainfall over the sea that an effective map of rainfall over the whole globe is still out of reach; nevertheless, what we have shows some features of interest and importance. July is a month of monsoon rains in India and on the West African coast, and there is also a good deal of rainfall over the Amazon valley, but not the prodigious quantity which occurs in April. It is a nearly rainless month in the Mediterranean region while in these favoured islands the rainfall differs little from that of any other month.

The average winds over the sea

Next, the distribution of winds; these, on the contrary, are for the sea only (Figs. 72 and 73); this circumstance does not compensate us for the inadequacy of the previous map, but adds to the feeling of vexation to which these gaps in our information give rise. There are reasons for both of them; but I can only apply to them the brusque expression which a former registrary once threw at me when, as praelector, I was late with a batch of supplicats for what I thought was a very good reason. "A very poor excuse," he called it.

The distribution of pressure and its relation to wind

We can make up in large measure for our lack of summaries of wind by attention to the distribution of pressure over the surface which appears in the next illustrations (Figs. 28–31, plates XX–XXIII). Such maps of pressure for the globe show the normal isobars or lines of equal pressure for sea-level; the first set was produced by Dr Alexander Buchan. They mark the first definite stage in our comprehension of the ways of the air because they illustrate the first inductive principle of modern meteorology, Buys Ballot's law, which explains that we ought to regard wind as air flowing along the lines of the isobars with the low pressure on the left (in the Northern hemisphere) and

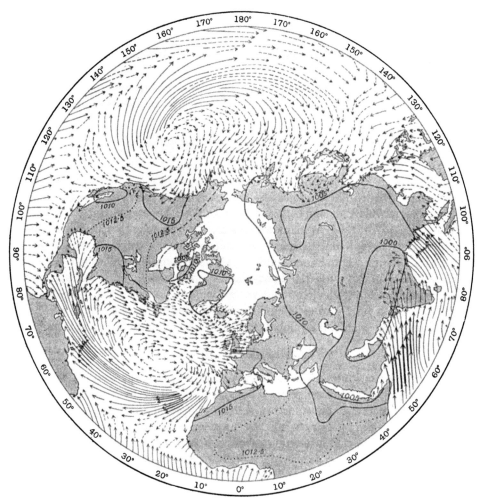

Fig. 72. Winds over the Oceans (Northern hemisphere), July and August,
together with isobars for July over the land. (After Köppen.)

The continuous arrows ⟶ ⟶ ⟶ denote steady winds.
The short arrows ⟶ ⟶ ⟶ ⟶ denote variable winds whose prevalent direction is that of
the arrow:
Wind Force, o—12: ⟶ under $3\frac{1}{2}$, ⟶ $3\frac{1}{2}$ to $4\frac{3}{4}$, ⟶ $4\frac{3}{4}$ to 6, ⟶ over 6.

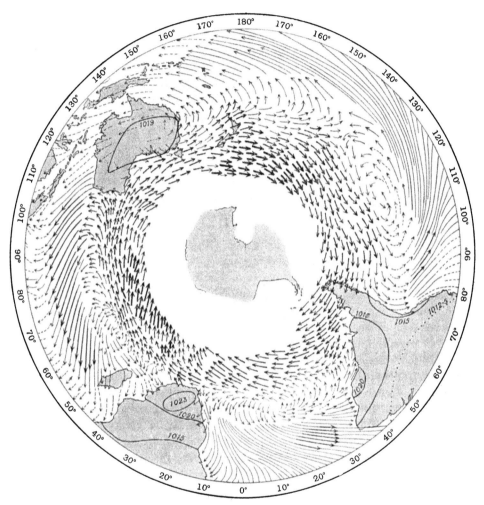

Fig. 73. Winds over the Oceans (Southern hemisphere), July and August,
together with isobars for July over the land. (After Köppen.)

The continuous arrows ⟶ ⟶ ⟶ denote steady winds.

The short arrows ➡ ⟶ → ⇢ denote variable winds whose prevalent direction is that of
the arrow:

Wind Force, 0—12: ⇢ under $3\frac{1}{2}$, ⟶ $3\frac{1}{2}$ to $4\frac{3}{4}$, ⟶ $4\frac{3}{4}$ to 6, ➡ over 6.

more or less deviation across the isobars from the high pressure to the low. We have learned to attribute the deviation to the obstruction caused by the surface and to add that in any locality the winds are stronger when the isobars are close together. We go so far as to assume a numerical relation between the spacing and curvature of isobars and the wind in the free air. These conclusions drawn originally from synchronous charts representing pressure and winds for one point of time are true enough of the average maps and therefore a map of isobars gives us also the distribution of winds at sea-level. It may be objected that wind at sea-level has little meaning in the Alps or the Himalaya, and that is true; but we go a stage further and conclude that at each level the winds are the expression of the distribution of pressure at that level and so we can bring the upper winds into calculation when we are able to compute or observe the corresponding distribution of pressure.

4. The Recent Investigation of the Upper Air

This remark may serve as the gateway to the next section of my subject, the information which has been collected in recent years about the upper air. Systematic investigation began towards the end of last century with observations in free balloons in Germany and observations with kites at the late Mr Lawrence Rotch's private observatory at Blue Hill, Massachusetts, now an establishment of Harvard University. Observations with ballons-sondes, that is, with unmanned balloons carrying self-recording instruments, were begun by Hermite and Besançon in France and developed by Teisserenc de Bort in that country, and by Assmann and Hergesell in Germany. The investigation was fostered by an International Commission appointed at Paris in 1896, and has been taken up in France, Germany, Great Britain, Russia, Italy, Bavaria, the United States, Canada, Java and Australia, with occasional contributions derived from expeditions in Scandinavia, the Atlantic Ocean, Victoria Nyanza, the Indian Ocean, McMurdo Sound (Antarctic), Spitsbergen and Greenland. Our own contribution has been mainly due, so far as measurements of temperature and pressure are concerned, to Mr W. H. Dines, a member of this University, whose work has not yet been sufficiently acknowledged. He began with kites at his own house at Oxshott under the auspices of the British Association and the Royal Meteorological Society in 1901, and received a grant-in-aid from the Meteorological Office from 1905 to 1915, when the work became a recognised part of official duty. So far as measurements of the upper winds with pilot balloons are concerned the initiative in this country is due to another member of this University, Mr C. J. P. Cave, of Ditcham Park, who responded to a letter which I wrote to *The Times* in 1903. He put his experiences with pilot balloons together in a book published by the University Press in 1912[1]. About that time the work upon pilot balloons was taken up officially for the Meteorological Office, the Royal Flying Corps and the Advisory Committee for Aeronautics by Mr J. S. Dines and Mr G. M. B. Dobson, and is now extended to the official aerodromes. Observations of the

---

[1] *The Structure of the Atmosphere in Clear Weather.*

smoke produced by bursting shells from anti-aircraft guns on a plan devised by Professor A. V. Hill were much used during the war, thus putting into practice a suggestion made by Sir Francis Galton as long ago as 1879. Occasional observations of pressure and temperature are now made by aeroplane, and observations with kite balloons are in process of inauguration.

The apparatus which Mr Dines devised for measuring the pressure and temperature with sounding-balloons is remarkably compact and portable. That, with the equipment for observing pilot balloons which is now reduced to standard patterns, is placed on the table for your inspection.

The heights reached by ballons-sondes are extraordinary. They are indicated in Fig. 74. Thirty-five kilometres (22 miles) is claimed as the record height.

The results of this investigation which I shall show you are firstly photographs of models of the structure of the upper air from Mr Cave's book, in which arrows flying with the wind show the average velocity and direction for each kilometre of height. I have selected two out of his six types for a particular reason, viz. that shown in Fig. 99 and another of exactly similar shape representing velocities in a current from the opposite direction.

The distribution of temperature in the upper air

My next diagram places Mr Dines's results for the temperature and pressure up to 20 kilometres over England in juxtaposition with those obtained in other countries so arranged as to show a section of the atmosphere (irrespective of hemisphere) from the summer pole to the winter pole (see Fig. 55).

5. A MODEL OF THE NORMAL ISOTHERMAL SURFACES

The outstanding feature of this investigation is the discovery of an upper region called the stratosphere where the temperature of the air is arranged in vertical columns and a lower region, called the troposphere, where the temperature is arranged in horizontal layers, with a surface marking the boundary between the two which is called the tropopause. This structure can be illustrated by a model of the surfaces of equal temperature (Fig. 47) which shows that all temperatures from the highest to the lowest exist over the equator and the surfaces of equal temperature which proceed therefrom, pass along nearly horizontally for some distance and are divided according to their ultimate course into three groups. Those of the lowest group come to an end at the earth's surface and mark the distribution of temperature at the surface which we have seen already. Another group carry on their horizontal course all round the world, while those of the third group pass along nearly horizontally until they meet the tropopause and suddenly take a turn vertically upwards. So it comes about that starting from the hottest region of the earth at the equator and travelling along the surface we find the air getting colder and colder towards the polar region, then a short journey in the vertical brings us to the stratosphere and then moving back again along the tropopause within the stratosphere, from the polar to the equatorial regions we raise

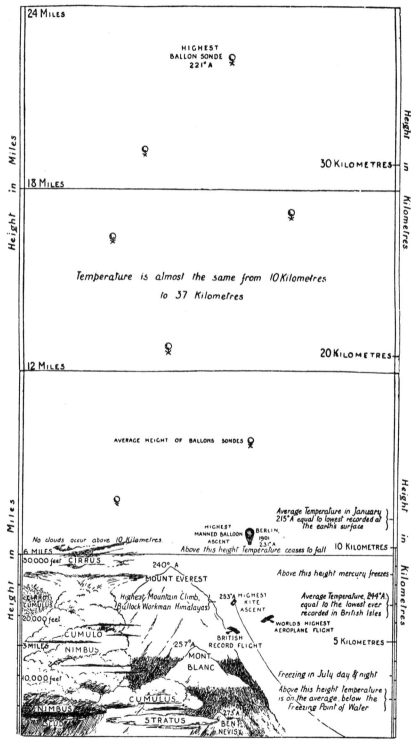

Fig. 74. Clouds and mountains in the troposphere; above them the region
of the stratosphere explored by sounding balloons.

our position a little, getting gradually higher and also gradually colder, until the lowest temperature of the world is reached when we are once more over the equator. Hence, the arrangement of temperature which we are accustomed to associate with the surface is completely reversed up aloft.

Cyclones and anticyclones

The model opens up to us a part of the subject which has hitherto been automatically excluded from what I have presented to you because I have used mean values or normals for my maps. I mean the phenomena of travelling cyclones and anticyclones. These occur especially in certain regions of the earth. Permanent anticyclones are shown on the map of pressure, permanent cyclones would also be shown if there were any but there are not; of travelling cyclones there are plenty but they are not shown, because though they frequent recognised paths all the localities along the path are visited by them in turn. Hence the regions of low pressure on the map of normal pressure must not be interpreted as permanent cyclones, but as regions across which cyclones or cyclonic depressions pass more frequently than elsewhere.

To deal with travelling cyclones and anticyclones we must use separate maps for separate hours; but the model shows how cyclones and anticyclones may be fitted into the map of normal distribution of temperature. The tropopause which marks the transition from troposphere to stratosphere comes down low in the region of a cyclonic depression and the surfaces of equal temperature meet it and are prematurely turned into the vertical, whereas over an anticyclone the tropopause bulges upward and the isothermal surfaces miss it and pass onwards further towards the pole while a sort of detached portion of the equatorial region constitutes itself overhead.

So it comes about that a cyclone in which, contrary to the older view, the air is colder in the troposphere than in an anticyclone becomes warmer than an anticyclone in the stratosphere because the stratosphere comes down to meet it, and *ipso facto* gets warmed[1], and so we find that the differences of pressure which we observe at the surface are almost fully expressed by the difference in weight of the columns of the stratosphere far above our heads from eight kilometres upwards.

6. The Approach to the General Problem of the Atmosphere

So much for the collection of facts about the air; you will notice that they are very numerous and very various and I hope you will agree that if those were the only result of the study of the air, the study is not entirely without interest. My final claim on your patience is an endeavour to show the possibility of finding some connection between all these facts on the basis of the laws and principles of physics. Here I ought to point out that for a long time past the science of physics has been able to supply explanations of isolated phenomena such as the colour of the sky, rainbows, haloes, the physical

[1] *The Free Atmosphere in the Region of the British Isles: A Note on the Perturbations of the Stratosphere.* M.O. Publication No. 202, 1909.

conditions of the formation of rain, hail, perhaps also of lightning and thunder, and of the separation of the atmosphere into troposphere and stratosphere. What has not yet been satisfactorily explained is the normal distribution of pressure and the average winds accompanying it and also the frequent disturbances by travelling cyclones and anticyclones. The whole succession of atmospheric phenomena is necessarily connected as a dynamical sequence and there are certain philosophers now at work who maintain that, given the present state of the weather with suitable exactitude, the ways of the air are so regulated by the laws of dynamics that the future cannot escape the process of rigorous computation; and by parity of reasoning, given a representation of the present conditions with approximate exactitude. the weather cannot escape approximate computation. The computation would be laborious; it has been jestingly said that, given the specification of the conditions to-day, it might take a year for the computers to work out a forecast of to-morrow and that would be too late. I am not out to say whether or not the ultimate solution of the interminable problem of the sequence of weather will be found in voluminous computation; but I want, for my own satisfaction, to generalise our conceptions of what is always going on.

7. The Atmosphere a great Steam-Engine

I suppose we are all agreed that the atmosphere is in reality a great engine, partly an air-engine but more effectively a steam-engine, or at least a moist-air engine. Now the essential parts of a steam-engine, as we all know, are a boiler to supply it with heat, a condenser at a lower temperature to absorb the surplus heat and a fly-wheel to maintain the continuity and uniformity of its action. We describe the action of the engine as taking a supply of heat from the boiler, giving out heat to the condenser, and converting into work, useful or otherwise, the difference between the heat taken in and that given out. The mechanical work of an ordinary engine may include the lifting of water or the maintenance of the speed of a train, a ship or a mill, against the perpetual inroads of the forces, frictional or other, which would otherwise soon annihilate its motion.

Can we rightly use such language about the atmosphere and usefully contemplate the ways of the air from that point of view? I think we can, though the analysis of the phenomena from that point of view is difficult. The boiler is certainly there, I have shown it to you in the distribution of temperature with the great warmth of the equatorial regions. In the map of the distribution of water-vapour I have shown you where the steam is raised. The condenser is there also, partly in the shape of the vast cooling surfaces of the high lands of the arctic and antarctic regions, and of snow-covered mountains generally; but perhaps more effectively in the upper air, by radiation through the stratosphere which at a temperature of 190 a to 240 a (from 60 to 150 Fahrenheit degrees below freezing) is cold enough to be transparent, and must be regarded as allowing the loss of heat by radiation into space, while its resilience enables it to act like a piston upon the working substance beneath.

8. The Fly-Wheel of the Atmospheric Engine

And what of the fly-wheel and the work done by the engine? Surely the winds, whether of the general circulation or of the local circulation of cyclonic depressions are a fair representation of the fly-wheel. At the risk of laying myself open to the unpardonable sin of punning, which would be out of place on this serious occasion, I will point out that the fly-wheel is of unspeakable importance to flying because the flier can either attach himself to it and get carried along with it or he may have to labour to make headway against it. The choice of these two alternatives depends upon the airman's knowledge of its habits and behaviour, of its ways, in fact. The constituent parts of the fly-wheel at any time are the natural air-ways of the world. It was by hanging on to one part of the fly-wheel in the fifteenth century that Columbus discovered America, and by the aid of another portion, just two years ago (June 14, 1919) Sir John Alcock crossed the Atlantic in $16\frac{1}{2}$ hours, and on July 13 of the same year Air-Commodore Maitland landed R 34 at Pulham after a journey from New York in 3 days and 3 hours.

Its total energy is tremendous. According to my computation a certain small depression had kinetic energy of the order of 10^{24} ergs, and I make a guess of the whole energy of the motion of wind as something near 20 billion horsepower-hours. One half, the part that belongs to the equatorial belt, is the energy of a wheel of steel two metres thick, 4000 kilometres wide, 12,000 kilometres in diameter and making at least one revolution in 45 days.

The energy is constantly being reduced by its internal friction and by the opposition offered by the obstacles of the surface, the effects of which are part of the work done. Some of the work is beneficent as in the distribution of rainfall and snowfall which are indispensable for life on the planet, the water power of lakes and rivers, the turning of windmills and the driving of ships. Some of it is destructive as the violence of hurricanes and tornadoes, some of it has uncomfortable aspects like the waves of the sea.

The polar and equatorial circulations in the upper air as parts of the fly-wheel

One of the immediate results of the thermal operations is to maintain the great fly-wheel or to start new sections of it in the form of local cyclonic circulations. Omitting those for the moment, I want to put before you some information about that part of the fly-wheel which is expressed by the general circulation. We can do so by distinguishing and ultimately isolating those portions of the atmosphere which represent permanent parts of the general circulation. Our best method of procedure is by way of maps of pressure. We can compute the distribution of pressure for successive levels and verify the computation by the observation of pressure at the various points of observation. We can thence calculate the winds which correspond therewith in accordance with the general principle of the relation of pressure to wind,

THE CIRCULATION IN THE UPPER AIR

W. 0° E.

Fig. 75. Relief globe and isobars showing the circulation of the Atlantic anticyclone and the cyclonic circulation of
the south-west monsoon and a weaker
circulation round Greenland.

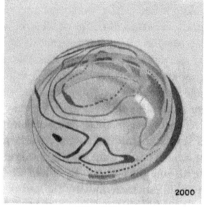

W. 0° E.

Fig. 76. (2000 metres.) Composite cyclonic whirl including a polar section and a
monsoon extension. Anticyclonic circulation over Africa and over the Atlantic
with extension towards U.S.A.

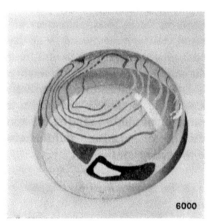

W. 0° E.

Fig. 77. (6000 metres.) Circumpolar
whirl becoming dominant. Monsoon
circulation hardly noticeable. Anticyclonic circulation over Africa and over the
West Indies and Southern U.S.A.

W. 0° E.

Fig. 78. (10,000 metres.) Vigorous
circumpolar whirl and anticyclonic circulation over Africa (with extension
over India) and over Southern U.S.A.

Isobars are drawn for steps of 5 mb. and in the charts of the upper air the band for each isobar
spreads over the area covered by a step of 1 mb. Where the band widens the gradient of
pressure is weaker.
The line across the bottom margin of the frame points to the meridian of Greenwich.
The arrows in Figs. 75 and 78 fly with the wind of the free air.
The figures reproduce photographs of a globe, 12 inches in diameter, and of three hemi-
spherical glass-shades 14 inches, 15 inches and 16 inches respectively in diameter, upon the
inner surface of which the isobars are painted.

to which reference has already been made and which finds partial expression in Buys Ballot's law.

I show you a selection of maps of the Northern hemisphere and if possible of models which express the results more clearly. The maps refer to the ground level and 8000 metres (Fig. 29, pl. XXI, and Fig. 61, pl. XVIII). The models (Figs. 75, 76, 77, 78) show the pressure at 2000, 6000 and 10,000 metres. They disclose an enormous body of air extending from the pole to latitude 40° or thereabout, circulating about the polar axis in curves not exactly coincident with circles of latitude but not very different therefrom. This mass of moving air is a very considerable fly-wheel. A reference to Fig. 63 (p. 130) confirms

Fig. 79. The " pole " of the summer monsoon—Mount Everest.

that view by disclosing the velocity of the winds at different levels for latitude 50° where the winds are near the maximum. You will notice the general similarity of the curve to Fig. 99 representing Mr Cave's observations with pilot balloons. The shape is evidently typical of the ways of the air. The collection of maps also discloses a collection of anticyclonic circulations in the intertropical region lying between a stream of westward moving air at the equator and a stream of eastward moving air at about latitude 35°. Thus the margins of the anticyclones form a sort of chain-drive pulling the air from east to west on the equator and pushing the polar circulation eastward. These vast local areas of high pressure are also interesting in relation to the tracks of hurricanes, which appear to circumnavigate the western sides of certain sections. The belts which separate the high pressure areas are over the coast-lines and emphasise the meteorological importance of those lines. With one of them the hurricane track of the West Indies is evidently associated.

In illustration of the winds which constitute the equatorial portion of the fly-wheel I show the results of observations of air-currents at Batavia, in Java close to the equator, carried out by means of pilot balloons (Fig. 64, p. 132). The general shape of the curve is singularly like that which we have already seen as expressing the computed winds over England and the two individual cases from Captain Cave's collection. There is shown, however, a westerly wind at 20 km. which is rather remarkable and is at present unexplored and unexplained.

The summer monsoon

Permit me here to call attention to the curious protuberance southward at the eastern side of the map for the level of 2000 metres, that marks the summer monsoon of India. It is much more conspicuous on the surface-map. The monsoon appears on these maps as a circulation in the lower levels round the high Himalaya and it would appear that the circulation near the surface in that region practically replaces the circulation round the pole which appears in the maps of higher levels. The air of the lowest 3000 metres has mistaken the Himalaya for the pole under the influence presumably of the cold on the northern side of those heights and has set up a circulation in defiance of the ordinary laws of latitude. It is not surprising when we consider the vast cooling influence which the mass of Himalaya's snow must wield in the summer months (Fig. 79).

9. Local Cyclonic Circulations as Parts of the Fly-Wheel

Among the products of the working of the aerial engine we have included the energy of the circulation of local cyclonic depressions whether they take the form of the hurricanes of tropical countries or the milder depressions of our own latitudes. I anticipate no objection to the suggestion that these phenomena are part of the working of the general atmospheric engine; but there is so far no general agreement as to the precise way in which the engine operates to produce these results.

I have recently suggested that the development of a vortex of revolving fluid may be due to the "injector-effect" or as I prefer to call it, "the eviction-effect," of rising air or falling rain or both combined, and I have put together an apparatus[1] designed to test the effect of the various possible causes in producing a cyclonic vortex when the conditions of relative motion are favourable. I find that cyclonic circulation may be produced by the exhaustion of air from the top or bottom of my enclosure, by the action of a jet of air passing upward or by the simple action of sand falling downward into a well beneath. I have consequently come to the conclusion that the air is much more easily moved to take up cyclonic circulation than has hitherto been supposed; and in fact cyclonic circulation is the natural expression of

[1] For detailed description of the apparatus see p. 103.

a part of the kinetic energy of rising air or falling rain, requiring only favourable local conditions for its obvious manifestation. Perhaps I may add that on that ground a volcano in explosive eruption ought naturally to cause a local tornado. I add, therefore, the energy of cyclonic motion to the other parts of the atmosphere's fly-wheel with some confidence in its accordance with natural fact.

10. AN INDICATOR DIAGRAM FOR THE ATMOSPHERIC ENGINE

Let us then summarise the working of the great steam-engine. In some parts it lifts air and water into the heavens and drops the water again as rain and thus provides the water power of the world. In other parts, particularly in the coldest parts of high lands, notwithstanding what I said about the peculiarity of the ways of air, cooled air descends. On the upward journey part of its energy is converted into the energy of cyclones and on its downward journey it maintains the energy of the general circulation. The process of developing cyclonic motion is helped by the falling rain and so the whole of the immense energy of the horizontal motion of winds is maintained or recreated.

If this view of the atmosphere is a reasonable one then we ought to be able to refer the operations of the air to what Maxwell calls an *indicator diagram* (Fig. 80), which shall express by the area of a closed figure the work done by the air in the course of the cycle of operations represented by the outline of the figure. I have done a number of other things during the past forty years; but off and on during that period I have been trying to get that diagram in continuation of work that I used to do

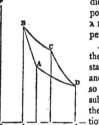

Let us suppose that the working substance is at the temperature T of the cold body B, and that its volume and pressure are represented in the indicator diagram by O *a* and *a* A, the point A being on the isothermal line A D corresponding to the lower temperature T.

First Operation.—We now place the cylinder on the non-conducting stand C, so that no heat can escape, and we then force the piston down, so as to diminish the volume of the substance. As no heat can escape, the temperature rises, and the relation between volume and pressure at any instant will be expressed by the pencil tracing the adiabatic line A B.

We continue this process till the temperature has risen to S, that of the hot body A. During this process we have expended an amount of work on the substance which is represented by the area A B *b a*. If work is reckoned negative when it is spent on the substance, we must regard that employed in this first operation as negative.

Fig. 80. Facsimile from Maxwell's *Heat*.

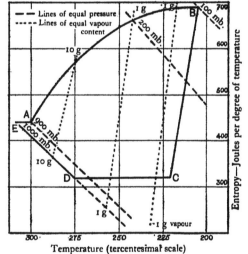

Fig. 81. Lines of pressure of one kilogramme of dry air saturated with water-vapour, and of amount of water-vapour contained therein, referred to temperature and "realised" entropy as co-ordinates.

with a class at the Cavendish Laboratory, and now I believe I have succeeded, with the assistance of Mr E. V. Newnham, of the Meteorological Office. The general result is shown in Fig. 50, and a suggestion of the cycle of operations in the atmosphere is given by the closed figure *ABCDEA* in my final illustration (Fig. 81). It is not exactly in the form which is familiar to readers of Maxwell, but in the form of an entropy-temperature diagram which Sir Alfred Ewing used in his work on the steam-engine. If it takes you as long to understand it as it has taken me to get it made, it is as obvious as, I hope, all the rest of my statements have been, that we shall not finish in time for lunch. Let me, therefore, content myself with saying that the air may be supposed to start its upward journey from *A* in a state of saturation and proceed speedily to *B* by its natural buoyancy. From *B* it may drift along the tropopause losing more heat by radiation than it gains and getting a little warmer in spite of the loss until it gets to *C* and has become cold enough to make an isentropic descent to *D*. From *D* to *E* it finds its course to the sea taking in heat and moisture and at *E* it has to make an isentropic atmosphere for itself in order to reach *A* once more and complete the cycle. Thus with the diagram it ought to be possible to make a reasonable diagnosis of the ways by which air can ascend from the surface and descending again thereto is prepared for a repetition of its cycle. We should thus replace by reason the guess-work which has hitherto done duty for it; further, according to the diagram the best which you can expect out of the steam-laden air of the equatorial region, working between the surface and the stratosphere under most favourable conditions is a "brake-horse-power-efficiency" of 25 per cent. Operations conducted elsewhere will have less efficiency than that. On the whole it is not very high, but the energy available as indicated by the equivalent of the amount of rain which falls is so enormous that there is no reason to doubt the capacity of the air as a steam-engine to develop and maintain the effects which are included in all our manifold experience of the air and its ways.

10. THE ORGANISATION OF THE METEOROLOGICAL OFFICE, IN LONDON, WITH SPECIAL REFERENCE TO AGRICULTURAL METEOROLOGY

A MEMORANDUM DRAWN UP AT THE REQUEST OF THE FOREIGN OFFICE, DATED 2 FEBRUARY, 1914, FOR THE USE OF THE FRENCH AMBASSADOR

HIS EXCELLENCY asked to be informed of the "*textes législatifs et administratifs* qui réglementent les services de météorologie générale et plus particulièrement de météorologie agricole en Grande Bretagne."

Textes législatifs. The only legislative authority for the Meteorological Services of this country in any year is the appropriation by Parliament of a sum as a "Grant-in-Aid" for the "Meteorological Office" in the Appropriation Act of that year.

A "Grant-in-Aid" is a fixed sum handed over by H.M. Treasury to be administered, under conditions laid down by the Treasury, by some body of persons, whether constituted expressly for that purpose or not, who become responsible for the expenditure, and for any administrative action taken in conformity with the prescribed conditions.

The Ministers responsible to Parliament for the grant, the Lords of the Treasury, accept no responsibility for the actions of the administrative body, provided they are within the prescribed conditions.

The grant for the expenses of the Meteorological Office was fixed at £20,000 in the year 1913. It is included in the Votes for Scientific Investigation, and is made to the Meteorological Committee, a body constituted by a Minute of the Treasury.

By custom, the Committee which administers a "Grant-in-Aid" is allowed to undertake the administration of other grants, and also to receive payment for special duties undertaken for or services rendered to private persons, public bodies, or, in certain cases, to Departments of Government. It is not entitled to the official services of the Post Office or other Departments of State, but by special arrangement the Meteorological Office enjoys certain official facilities with regard to the priority of meteorological telegrams and with regard to stationery and printing.

The total expenditure on the various services in the administration of the Meteorological Committee is about £30,000 a year. It must, however, be pointed out that the meteorological observations available at the Office include those which are carried on primarily for their own purposes and at their own expense, by municipal corporations and private persons. With regard to these, the Office acts only as adviser, organiser, compiler and publisher of results. This voluntary work—taking observations of rainfall also into account —probably represents an expenditure of £20,000, making the whole expenditure on meteorological services in this country, including municipal and private enterprise, about £50,000, of which £21,000 is provided by Government.

Textes administratifs. The only *textes administratifs* for the Meteorological Services are the Minute of the Lords Commissioners of H.M. Treasury, dated 20 May, 1905, constituting a Director of the Meteorological Office and a Meteorological Committee with the Director as Chairman, and subsequent Treasury Minutes re-constituting the Committee or appointing members thereof.

In accordance with regulation, accounts of the receipts and expenditure of the Office for each year are audited by the Comptroller and Auditor-General and reported to Parliament; and a Report upon the work of the Office is presented each year to the Treasury and laid before Parliament by Command of His Majesty.

Any Member of Parliament is, therefore, at liberty to raise any question upon the accounts or the report, but otherwise, with the limitations herein indicated, the Director and Committee have full discretion as to the objects to which the funds shall be devoted and the means which shall be adopted for securing them.

The practice of the Office is guided by tradition, which has been formed in the past 60 years. The Grant-in-Aid has been increased from time to time for reasons urged either by the controlling authority or by parliamentary critics of the Office. Each increase has carried with it the intention to accomplish some specific object, and therefore, a tacit obligation on the part of the controlling authority: but the Treasury has never made conditions about details of expenditure, and has always accepted the statement of the proposed allocation of the grant without comment; so that the Committee is not bound by any conditions, but merely guided by its own judgment in accordance with tradition and practice.

It is important to note this in the consideration of the special application of meteorology to agriculture. That is one of the objects of the Office, but any other of the applications of meteorology in the interest of the public is equally so. There is no special allocation of funds for the application of meteorology to agriculture as such.

The operations of the Meteorological Office being guided so largely by tradition and practice, without any *textes administratifs*, which prescribe its duties and operations in detail, the only means of obtaining a conspectus of the guiding principles of its organisation is by reference to the Annual Reports which have been issued since 1868, to occasional reports before that date and to the reports of certain Committees of Inquiry, from which the present organisation has grown. It now comprises a Central Office with four technical divisions, two branch offices, six meteorological observatories, about thirty subsidised stations and upwards of three hundred voluntary stations. Terrestrial magnetism and seismology are associated with meteorology. The reports are, however, so voluminous that a brief historical retrospect may be acceptable, and is therefore given here. It is necessary to go into some detail because the subject ultimately under consideration is the application of meteorology to agriculture, and in such a case application means that the agriculturist must be in a position to receive and use the information which the meteorologist has to give, and this requires much preliminary work, first, to put the meteorologist in possession of the necessary facts and principles, and secondly, to enable the agriculturist to understand the technical language and ideas without which communication is meaningless.

Marine Meteorology

1. The Office was started in 1854 as the Meteorological Department of the Board of Trade, under Admiral R. FitzRoy, on the recommendation of an International Maritime Conference held at Brussels in 1853. The sole purpose of the department was the supply of meteorological instruments to the navy and mercantile marine, and the collection and discussion of meteorological observations from ships.

Those duties still remain in much the same form. They are undertaken partly by the Instruments Division and partly by the Marine Division of the Office.

Daily Weather Study. Telegraphic Reporting and Anemograph Stations

2. Moved especially by the loss of the "Royal Charter" in 1859, Admiral FitzRoy in co-operation with Le Verrier, and with the support of the Prince Consort, began daily telegraphic reports from stations in this country, and subsequently from France. With the aid of charts prepared from the observations, he commenced the issue of "forecasts" and "storm-warnings." This was the first beginning of what is now the Forecast Division of the Office. The procedure was sharply criticised by the scientific authorities of the time, and upon FitzRoy's untimely death in 1865 the Board of Trade took the matter up with the Royal Society. Upon the report of an inter-departmental committee (Board of Trade and Admiralty) a new departure was taken, viz.:

Observatories of the First Order

3. In 1867 a grant-in-aid of £10,000 was assigned to a Meteorological Committee (unpaid) to be appointed by the Royal Society, with enlarged duties, viz.:

(a) To continue the work in marine meteorology of the Meteorological Department of the Board of Trade.

(b) To continue the study of weather by means of daily telegraphic reports, but not to issue forecasts or storm-warnings. (The storm-warnings were replaced by request of the Board of Trade, but the forecasts remained in abeyance till 1879.)

(c) To bring to the assistance of the study of weather, the records obtained from self-recording instruments at seven special observatories on land, established for the purpose. (This marks the introduction of observatories into the Meteorological Office system which are now represented by five meteorological observatories, two of which include magnetism and seismology, and a central observatory for the study of the upper air.)

Climatology and the Meteorology of the Globe

4. In 1872 an International Conference of Meteorologists was held at Leipzig, which was followed by official international congresses of duly accredited representatives at Vienna in 1873 and ultimately at Rome in 1879. These international meetings concerned themselves partly with the exchange of information by telegraph between countries in Europe, and also with the study of climate, which is primarily of local or national importance, but ultimately has to do with the meteorology of the globe.

Exchange of Publications

They also led to the organisation of an elaborate exchange of publications, so that a meteorological office has become the most cosmopolitan of all institutions and is in direct communication with every civilised country.

The international meetings not only brought an accession to the importance of the daily weather exchange, but introduced a new subject, climatology, into the work of the Office, which in England and Scotland had been the care of voluntary societies.

Thereupon the Scottish Society, through representatives in Parliament, demanded a subvention, and failing that, an inquiry into the administration. There was also much dissatisfaction about the marine work, and the inquiry was granted by the Treasury, who appointed a committee of inquiry under the Chairmanship of Sir William Stirling Maxwell.

Statistical Division. Special Researches

5. Upon the report of that Committee in 1877 the Treasury decided to revise the constitution, to place the actual direction as well as the general control of the Office in the hands of a paid Council appointed by the Royal Society with the sanction of the Treasury. The grant became £14,500. Climatological work was added to the obligations of the Office, and is now represented by the Statistical Division of the Office, which concerns itself with the publication of the official Year Book. Also by the same instrument "special research," which included experimental work of various kinds, was recognised as a legitimate object of expenditure.

Daily Information for Newspapers. Evening Telegraph Service

6. In 1879 the study of weather with the aid of daily telegrams and of self-recording instruments at observatories was pronounced to be sufficiently far advanced to justify the issue of forecasts, and they were accordingly issued at 11 a.m. daily to public offices and newspapers gratis, and to "subscribers." But *The Times* newspaper desired also an evening issue that might be printed

in the morning paper. For some time the money necessary for the additional service was provided by *The Times* and subsequently by a syndicate of newspapers. Then the Government accepted, the obligation and increased the grant first by £500 and subsequently by £300 more on that account. The Office was accordingly charged with a new duty—the supply of weather information in the evening to newspapers. It is now associated with evening duty for the Admiralty and Military Air Stations.

7. From 1880 things went on without change for more than twenty years, but in 1903 the Scottish Members of Parliament again demanded an inquiry on account of the failure of the Scottish Society to obtain a subvention sufficient to maintain the Observatory on Ben Nevis from the grant made to the Meteorological Office. Another inquiry was set on foot by the Treasury, and a committee appointed under the Chairmanship of Sir Herbert Maxwell. This Committee reported in 1904, and resulted ultimately in the Treasury Minute of 20 May, 1905, already referred to.

Réseau Mondial. Library

In the meantime the library had become a most important matter, and the compilation of information about climate and weather in the various parts of the British Empire practically constituted a new department of activity. The pressure on the one hand of the International Meteorological Committee and on the other hand of the study of Solar Physics, has gradually led to the recognition of an obligation towards a *réseau mondial* as specially incumbent upon the Meteorological Office as the central institution of its kind for the Empire. This has become part of the duty of the Secretarial and Library Division of the Office.

Investigation of the Upper Air

Since the Meteorological Committee was constituted in 1905 many changes have supervened; telegrams from Iceland, wireless telegrams from ocean steamers, the air departments of the navy and the army, the absorption by the Office of the whole duty as regards climatology, previously discharged by the Societies with the aid of a subvention. The Office has also become the central institution for the meteorological investigation of the upper air. It has taken over the direct control of observatories which were previously under separate authorities, and this step has brought with it the responsibility of the Office for certain aspects of terrestrial magnetism, as well as of atmospheric electricity and seismology These changes have been associated with an increase of the grant to £20,000.

Colonial Observations

The Office has also become an advisory centre for the Colonies which have no separate meteorological organisation and it also assists the Meteorological Institutes of the Dominions in the selection and purchase of instruments.

British Rainfall Organization

Thus the horizon of the work of the Office, which was originally limited to the collection and discussion of observations from the sea, has now become very wide; still it does not include all the British meteorological interests. The important subject of rainfall in the British Isles is still the care of a private organisation—The British Rainfall Organization. The Meteorological Office makes no attempt at the detailed representation of rainfall, and only deals with rainfall as part of climatology[1].

When, therefore, the application of meteorology to agriculture is considered, so far as the Meteorological Office is concerned, anything which is dependent upon the detailed study of the distribution of rainfall is not necessarily included.

In a recent communication to the Treasury, the purposes which the Office keeps in view have been defined in the following terms:

"(i) The collection of observations from ships on all oceans, together with the discussion and publication of meteorological results, for the benefit of sailors, and as a contribution to the meteorology of the globe.

"(ii) The collection and publication of reports received by telegraph, and the issue of forecasts and storm-warnings based upon them.

"(iii) The maintenance of observatories and anemograph stations to furnish material for the scientific study of the phenomena of weather as exhibited on the daily charts, and the application of the study to the improvement of forecasting, and other purposes.

"(iv) The organisation and maintenance of a trustworthy public memory of the weather, which is available for reference at any time by all classes of the community, and which forms a basis for the study of the climatology of the United Kingdom in comparison with that of other countries, and in relation to agriculture, public health, and other public purposes; the discussion of the observations with a view to the definition of climatic factors for this country in comparison with others, and ultimately to the relationships of seasons and the establishment of more general laws of climate and weather that should lead up to a reasoned forecast of coming seasons.

"(v) Co-operation with the British Dominions and foreign countries for improving the organisation and the instruments by which the purposes enumerated above are to be pursued, and for the effective representation of the meteorology of the globe."

Apart from the question of special researches by individuals at the Central Office or at the Observatories, the means which are adopted by the Committee for securing these objects are set out in the Circular 001 of which a copy is annexed, together with a copy of the latest report of the Committee, which gives on pp. 6–8 the names of the staff, consisting of about 80 persons.

These facts will enable His Excellency to form an opinion as to the rather

---

[1] The British Rainfall Organization was transferred to the Meteorological Office in 1919, and the Office itself incorporated with the Air Ministry in the same year.

complicated structure which is represented by the meteorological organisation of this country. References are given to the original documents which form the material through which the gradual development of the structure can be traced. It consists of a central office, with branch offices, observatories, and stations. The work of the central office is in five divisions, viz.:

1. Marine Meteorology.
2. Forecasting, Storm-warnings, and Dynamical Meteorology
3. Climatology and Statistics.
4. Instruments and equipment for observatories and stations.
5. Library, Inquiries, and Réseau Mondial.

Meteorology and Agriculture

Some addition is necessary with regard to the important and difficult question of *la météorologie agricole*. It is really an open question whether the responsibility for the application of meteorology to agriculture belongs to the Meteorological Office or to the Board of Agriculture and Fisheries in England and the corresponding departments in Scotland and Ireland. The traditional attitude of the Meteorological Office is that it collects and digests meteorological information which the agriculturist can apply if he wishes, and from that point of view the following issues of the Meteorological Office are regarded as suitable.

Forecasts

1. The *Daily Weather Report* with the provisions set out for telegraphing forecasts for a small fee to those who are willing to pay for the telegrams. Circular 001 (A–H), p. 9.

Forecasts are prepared throughout the year each morning at 10 a.m. and each evening at 7 p.m., and during the harvest season—June to September—in the afternoon, specially for agriculturists. Circular 001 (A–H), p. 13.

Statistics

2. The *Weekly Weather Report*, which was projected especially with a view to agriculture and public health, gives a summary of the pressure, temperature, sunshine, and wind in a form which was designed to be especially suitable for agricultural purposes. This report has now been continued for 36 years, and forms a homogeneous body of statistics week by week which is, for that purpose, probably unrivalled in the world. But it has a very small circulation outside official circles.

3. The *Monthly Weather Report* which gives the usual climatological information for about 300 stations in the British Isles.

In actual practice these provisions are very little used by agriculturists. Many persons are willing to receive forecasts by telegraph but are unwilling to pay for the telegrams; it is entirely contrary to the instinct of the British race to pay for anything until its value has been made undeniably clear, so that the farmer and the Government are both waiting for the utility of the forecasts to be demonstrated beyond cavil. Yet that can only be done by trial, and nobody has yet been found who is willing to pay the cost of an adequate trial on a large scale. The Meteorological Office could, if the Committee wished, undertake that experiment, but it would mean diverting some of its funds from meteorological study to meteorological applications. It is naturally disposed to make quite sure of success before it embarks on a speculation of that kind, and certain success is the reward of careful study. No institution with scientific instincts is disposed to commit itself to the position that its knowledge is complete and that it can forego any further investigation, especially in such a subject as the study of weather.

The climatological aspect of *la météorologie agricole* is a matter of the greatest difficulty. The practical farmer has made his own study of weather and used it in his own way without committing the results to writing. The Meteorological Office commits a vast number of figures to print without knowing what their precise application to agriculture is. All are agreed that agriculture depends upon weather, but to ascertain the manner in which the figures of the meteorologist can be applied to supplement the farmer's practical experience of weather is a matter requiring something that approaches to genius.

The relations between the Meteorological Office and the Boards of Agriculture in the United Kingdom are of the happiest, but neither side knows exactly how or where to begin. Some progress has, however, been made in this country. Some years ago the Meteorological Office issued a note about the wheat crop in relation to the rainfall of the previous autumn, and this was taken up by a member of the staff of the Board of Agriculture, who produced a most valuable discussion by modern statistical methods of the relations of weather and crops for one district of England.

Education

The further development of the application of meteorology to agriculture is largely dependent upon education in the rural schools. The study of weather is now becoming a part of education in many schools, rural as well as urban, so that the prospect of more effective organisation is good. The provision for this is shown in Circular E. 03.

But thus far as regards organisation, at present the formal responsibility of the Office is limited to preparing forecasts, and compiling statistics which will be indispensable when further investigation has so far developed the laws of weather as to allow of forecasting coming seasons.

That is one of the avowed objects of the *réseau mondial*, and the work thereupon must therefore also be regarded as a contribution to *la météorologie agricole*, although the practical farmer would probably not so regard it.

Answers to Inquiries

Perhaps the most valuable provision of the Meteorological Office at the present stage is the provision for answering inquiries about the weather on the part of the general public. Any public department and any private person may ask any question that can be answered by a knowledge of the facts and laws of weather, and to such questions answers are given with all the intelligence that the Office can command. Many inquiries are answered, and the inquirer often finds the Office to be possessed of information of which he was unaware.

This provision allows inquiry to be directed along the lines which the agriculturist opens; among the subjects which have already been the subject of inquiry may be mentioned—spring frosts, and the protection of vegetation by "smudging"; autumn frosts; the effect of gun-fire upon rainfall, particularly during harvest; spells of fine weather for harvest; temperature in relation to sugar-growing; the limits of forestation prescribed by temperature; atmospheric humidity in relation to brewing.

By watching the trend of these inquiries, and by the organisation of the means of preparing intelligent replies, the Meteorological Office hopes to approach the question of *la météorologie agricole* on lines suggested by agriculturists themselves, and at the same time by encouraging the development of weather study in schools to lead up to the spontaneous use of the information compiled in the Office.

If necessary, the form of the information which meteorologists have hitherto put forward as representing the main features of climatology will be altered so as to meet the needs of the agricultural inquirer.

In fine, it may be said that at present the Meteorological Office is more concerned with the means for organising *la météorologie agricole* on a satisfactory basis than with any organisation actually in operation.

11. CLIMATOLOGICAL STATIONS AND LOCAL AUTHORITIES

A Memorandum approved by the Meteorological Committee[1] in 1915

In the present emergency in national affairs the Meteorological Committee desire to call attention to the position of the Meteorological Office in relation to the collection of observations from what are technically known as "climatological stations"; that is to say from stations which are maintained, not by the Office in connection with the public daily service of forecasts and gale warnings, but by local authorities or private persons. They contribute observations to be used by meteorologists for the study of the details of climate and weather in the British Isles, and by the public who require information about the weather for various purposes.

The Meteorological Office is a central depository of transcripts of meteorological observations of various kinds in every part of the British Isles, of the British Empire, and, indeed, of the whole world, not because the information is essentially necessary for, or immediately applicable to, the work of forecasting and the study of daily weather which, so far as observations on land are concerned, are its primary duties, but because an organised central storehouse or memory of the experiences of weather for a long series of years is of great public utility and more effective than any compilation which otherwise individuals would be able to make for their own use. By agreement between the office and the Scottish Meteorological Society, the Meteorological Office, Edinburgh, discharges a similar duty with special reference to Scotland.

In the course of the past 20 years, a large amount of valuable information has been compiled, the existence of which is hardly realised. It is still far from complete, but I may be permitted to illustrate the usefulness, or at least the appositeness, of an efficient public memory by recalling a report which I happened to see some years ago in *The Westminster Gazette* of a law-suit in which a tobacconist sued his neighbour for damage to a case of cigarettes, alleged to be due to rain coming through a broken skylight. It was acknowledged that the skylight was broken by the neighbour's son and, according to the report, "all went well until a mild-mannered gentleman from the Meteorological Office" proved that it had not rained since the skylight was broken, and the plaintiff's case had to be abandoned.

In order to be effective the collection of information should be carefully organised. The preservation of a trustworthy and sufficient memory of past weather is primarily a matter of urgent local importance. The weather is an element in the profit and loss account of every individual, of every parish,

[1] Tenth Annual Report of the Meteorological Committee to the Commissioners of His Majesty's Treasury for the year ended 31 March, 1915. Cd. 8028, pp. 75–78.

of every district, whether urban or rural, of every county and of every state; and the preservation of an efficient record of these events is just as important for the persons or authorities concerned as the record of the money transactions in which they are engaged. The difference between the two sets of experiences is that one is beyond the control of the individual or local authority and the other is not; but no steward of his own or other people's interests would be regarded as wise if he left out of account the gains and losses which he could not control.

The question of meteorological observations, or weather records, may be put in this general form: Here is a spell of rain, which the house-gutters, the local drains, the roads, gulleys and streams have to carry away, a snow-storm which may make the neighbourhood impassable, a hail-storm which damages the crops, a drought, or a long frost which endangers the water supply, a wind which brings down all the loose tiles and chimneys. Are these events to be regarded as normal and to be provided against by suitable precautions, or are they outstanding risks which should be left to chance?

Only by an adequate public memory can an answer be given, and hitherto the provision of the material for an answer has been left mostly to private enterprise. The claims of science have usually been urged as an encouragement to private enterprise, and without doubt, such observations are indispensable for the scientific study of weather; but they are equally indispensable for the proper conduct of the ordinary affairs of life. Since the study of weather began to be organised on a scientific basis, circumstances have changed. The life of the individual and of the community is not nearly so self-contained now as it used to be; it is much more dependent upon facilities for communication with the rest of the world; the increase of those facilities enables the experience of many to be used for the advantage of each in a far greater degree than was possible in the olden days.

To take an example, the practice of insurance is far more widely spread than it used to be. Taking the case of insurance against hail, the premium should be different according to the locality; but so far as is known, the localities in their corporate capacity keep no records, and in consequence the premium is fixed for them upon information privately compiled by the insurance companies—that is to say, by one of the parties to the bargain. Many other forms of insurance against weather are possible, but only when the risk can be properly computed by means of ascertained facts. This Office has recently been concerned in a legal dispute as to whether damage to property during a squall of wind accompanied by incessant lightning was directly due to the wind or to the lightning: a fine distinction, upon which the validity of the insurance turned, and which suggests some revision of the practice of insurance in the light of recorded experience of weather.

Local authorities have given little consideration to these matters, and individual farmers and others have trusted to their own reminiscences. It is, in fact, apparent that the balance of prosperity has been so large that it has not hitherto been felt necessary to pay much attention to the profits to be made out of the weather, or to economise the losses which it causes. But when the pinch of adversity comes, as it must come after the squandering of so much of the world's wealth in the war, the reduction of any risk by the use of organised knowledge is at least worthy of consideration. The stress of war is therefore a reason for organising the study of weather, not a reason for postponing organisation to a more prosperous season.

No one will deny that a careful record of the weather regularly compiled from day to day on a definite plan is in the long run a better basis of action than the longest stretch of personal reminiscences, just as a daily record of river-level is better than an occasional mark on the parapet of a bridge. With the change of circumstances, from the comparative independence of the homestead to mutual dependence of town and country, and from the abundant prosperity of past years to the adversity that lies in front of us in the near future, the preservation of an adequate record of the events of weather for comparison with past times and with other localities has also changed from being a matter of scientific and personal curiosity to a necessity for the community. It is from that point of view that it should be regarded; the additional advantage that may accrue from scientific meteorological study is all to the good, but it is another matter.

The unanimity with which the health resorts have made provision for careful records of weather show that a knowledge of the weather must be looked upon as a valuable asset, and it is equally so for any other locality. A contractor who undertakes work for a local authority must either know something about the weather or allow a wider margin for contingencies than is really necessary; the locality must either supply the information or provide the margin.

Hitherto the observations upon which we depend for supplying information about the weather in all parts of the British Isles have been largely those of country clergy and landowners; but the drain upon their resources, particularly in men, has begun to diminish the number of observations available. Already in Ireland the observations are altogether inadequate, and when, for example, questions are put as to the parts of the country where climatic conditions are favourable for afforestation, we cannot give a satisfactory answer, because the localities have made no record of their experience. Moreover the distribution of observing stations depends not upon the present and future requirements of the public, but upon the existence of a local volunteer.

It is submitted, therefore, that the Local Authorities should give serious consideration to the question of an adequate record of weather. The Meteorological Office has been active in collecting and organising the meteorological information that was known to be available. This has given the impression that the Office, as the creation of the central Government, ought itself to provide any observations that may be found necessary for any purpose whatever; but such an impression is quite erroneous. Out of 500 observatories and stations which contribute observations to the Office for the benefit of the public, only 36 are maintained or subsidised out of Office funds. A considerable number are maintained by local or statutory authorities and the remainder by private persons at their own expense. It is natural, and perhaps laudable, that in the matter of weather the City of Westminster should rely upon the Meteorological Office, instead of itself, for its memory; and it is not unreasonable that the Office of Works, in a dispute over a contract, should apply to the Office (unsuccessfully, I fear) for details of weather between Avonmouth and Bristol, and their relation to the average; but it would be absurd, for example, for the Council of the County of Warwick to rely upon London to know what weather had been experienced in Warwickshire; or for residents at Hindhead to live in ignorance of their own climatic conditions until the Government provides the information. The natural order is just the reverse; the Meteorological Office should naturally appeal to the localities to know what has transpired there, and it is a matter for surprise how many of the County Councils, when appealed to, would be unable to say what the weather had been in their county since it was under their charge. The whole situation arises from the mistaken notion that to satisfy the condition of utility at all, knowledge must be useful here and now, and that nothing need be preserved for which the officials of to-day have no obvious and immediate use. It is the memory which goes back longest that is the most effective, and therefore most useful.

It ought, in fact, to be the function of the Meteorological Office to reduce, rather than to multiply, meteorological observations, by proper organisation and by the suggestion of co-ordination, where co-ordination is economical. The following guiding principles seem to be applicable: for keeping its water-supply and drainage properly under observation, every parish ought to have its rain-gauge, and the Parish Council might see to that. A District Council might keep a regular record of temperature and weather as well, for its own district; while in every county there should be, for official purposes, a proper number, and no more, of fully equipped climatological stations which should be centres of information about the weather and its ways for all concerned.

12. THE LAW OF SEQUENCE IN THE YIELD OF WHEAT FOR EASTERN ENGLAND. 1885—1905[1]

REPRINTED FROM *The Journal of Agricultural Science*, vol. II, part 1, Jan. 1907

IN February, 1905, I made a short communication to the Royal Society of London (*Proceedings*, vol. LXXIV, p. 562), upon an apparent relation between the yield of wheat for England in the 21 successive years 1884 to 1904 and the aggregate rainfall for the three autumnal months (September, October, November) of the preceding years of the principal wheat-producing districts of Great Britain. The relation was put into the algebraical form:

$$\left.\begin{array}{l}\text{Yield of wheat for}\\ \text{England, per acre}\end{array}\right\} = 39\text{·}5 \text{ bushels} - \tfrac{5}{4}\left\{\begin{array}{c}\text{previous autumn rain-}\\ \text{fall in inches}\end{array}\right\} (1).$$

In considering the application of the formula to the computation of the yield for the 21 years specified, 7 years were found to give large differences, and reasons were given for regarding the 7 years as being otherwise anomalous.

An endeavour to pursue the investigation of this suggestive relation between yield of wheat and autumn rainfall has disclosed a curious relation between the numbers representing the yield of wheat for a selected part of England, during the last 21 years. This relation is irrespective of any connection with the previous autumn rainfall, but as it was disclosed in the course of the investigation of that connection and may ultimately prove to be associated therewith, I will set out the figures first from that point of view.

The formula already referred to in the *Proceedings* of the Royal Society was based on figures against which objection might be raised, on the ground that they were not strictly correlative. The wheat values were the averages for England; the rain values the averages for those districts of Great Britain where wheat is grown in considerable quantity. It is not easy to obtain figures which refer to an area sufficiently large to eliminate accidental local influences and are, at the same time, properly correlative; but I endeavoured to obtain correlative figures by taking the rainfall values for the district "England East" of the Weekly Weather Report and compiling the corresponding values of the yield of wheat from the results given in the official returns of the Board of Agriculture for the counties included in the district.

These figures are, so far as I know, quite unexceptionable for the purpose of the comparison in view. They are doubtless subject to incidental errors, but both on the side of rainfall and on that of yield they are homogeneous

---

[1] The earlier part of this paper is a reproduction with slight modifications of a paper on "The Law of Sequence in the Yield of Wheat for Eastern England, 1885–1904" contributed by the author to the "Hann Band" of the *Meteorologische Zeitschrift*. Vieweg und Sohn, Braunschweig, 1906.

for the period to which they refer. It is a matter of regret that they do not extend beyond 21 years, the limit set by the wheat returns. The figures for rainfall and for yield of wheat are given in a table at the end of this paper.

The subsisting relation between the autumn rainfall and the wheat-yield of the subsequent year is sufficiently evident, though the constants are different from those in the equation (1) already quoted. For the Eastern Counties the variation of yield in the 20 years is from 25·2 bushels per acre in 1893 and 1904 to 36·3 in 1898, and the formula

$$W = 46 - 2\cdot2R \qquad \qquad \ldots\ldots(2)$$

gives the yield within the limit of accuracy of 2·1 bushels per acre for 13 years out of the 21 including 1905; but there are eight exceptional years when the yield computed by the formula differs from the actual yield by more than 2·1 bushels. The exceptional years are not all the same as those which were exceptional in the comparison for the whole of England, and on two occasions in particular, viz. 1886 and 1903, the differences, one in excess, the other in defect, are very great.

In the previous comparison an explanation of the differences in exceptional years was given that was *primâ facie* reasonable, and I examined the figures for the East of England in the hope of finding a similar explanation for the differences. In such an inquiry one naturally regards the rainfall data as fundamental, and looks for an explanation of the deficient or too abundant harvests in the accidents that may occur to the crops from floods, hailstorms and other causes, because the damage done by such accidents bears no numerical proportion to the quantity registered by a rain-gauge. When I plotted diagrams representing the two quantities, rainfall and yield, I expected to find marked irregularities in the yield diagram, with no counterpart in the rain diagram; and I was, for that reason, the more surprised to find that the yield diagram exhibited a singular regularity of sequence, while it was the rain curve that had what may be called abnormal irregularities.

The most singular point about the regularity of sequence of the yield diagram is that the curve is almost perfectly reversible with reference to the epoch 1895–6. The average yield for the two years, 1895, 1896, is (31·1 bushels) almost exactly the same as the average yield for the 21 years (30·9 bushels); thus 1896 compensates for 1895, and in the same way 1897 compensates for 1894, 1898 for 1893, and so on, without any really considerable exception throughout the 21 years. The worst case of all, an exceptionally bad one, is that for 1887–1904, which, however, gives a mean differing by less than 2 bushels from the average for the 21 years.

Fig. 82 represents this curious coincidence. Curve *A* shows the yield of wheat for England East in successive years, and curve *B* shows the yield *for the same years taken in the reversed order*. It will be seen that curve *B* is very similar to the simple reflection of curve *A* from the base line. The numbers at the foot of the figure show the mean value of the ordinates of the two curves in bushels per acre. If the curve B were a perfect reflection of curve A, all the mean values would be identical, and it will be seen that they are very nearly so. They range between 29·1 for the pair of years 1887–1904, and 31·8 for the pair of years 1890–1901. The average for the whole 21 years is 30·9, so that the greatest difference of the mean of any of the pairs of years from the average of the 21 years is 1·8 bushels per acre in the one direction, and 0·9 in the other.

The reversal of the curves can be more strikingly exhibited if one of them be traced and the trace turned round and superposed on the original.

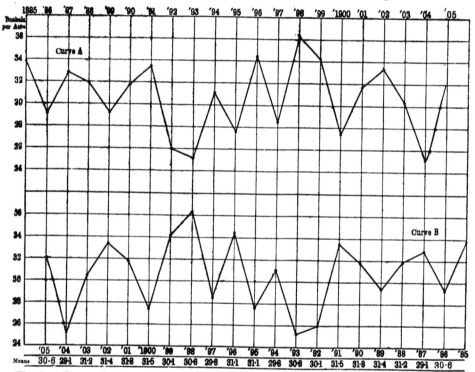

Fig. 82. Curve A: The yields of wheat 1885 to 1905. Curve B: The same yields in reversed order. The means of years equidistant from 1895–6 are shown in bottom line to be nearly equal throughout.

It seems scarcely possible that this compensation, persisting through so many years, should be fortuitous. It would be accounted for if the figures representing the yield of wheat were points on a periodic curve of complex periodicity such that a number of important components concurred in a nodal point in the epoch 1895–6. The test for a nodal point between 1895 and 1896 would be that the ordinates should be reversed in the way in which they are very approximately reversed in actual fact.

It is surprising that this evidence of periodicity exhibits itself in the wheat values, while it is certainly not conspicuous in the autumnal rainfall. The years that are exceptional as regards the rule of parallelism between the autumn rainfall and yield of wheat, viz. 1886, 1889, 1896, 1897, 1901, 1903, and 1905, do not appear as exceptions to the rule of reversal of the wheat yield with reference to 1895–6. The real value for the yield in 1903 is 6·2 bushels below the amount calculated from the autumn rainfall formula, but it compensates the yield for 1888 quite satisfactorily. The only year which shows considerable divergence from the rule of compensation is 1904, which agrees very well with the rainfall formula, but has too small a yield to compensate 1887. It will be seen from the sequel that the defect may probably be attributed to the 1887 yield.

Assuming from the reversal with reference to the epoch 1895–6 that the yield can be represented by a series of periodic components concurring in a node at that epoch, I have endeavoured by an examination of the figures to determine the period and amplitudes of the component oscillations.

There is, so far as I am aware, no organised method of doing this, except for the case of periodic variations of known period with harmonic components and a large number of ordinates for a complete period. In the present case the fundamental period could not be regarded as known. A careful examination seemed to show that the curve might contain components of 2 years', $3\frac{1}{2}$ years', and 11 years' period, and the combination would then only recur completely in 154 years.

To deal with complex periodic variations of unknown period we can proceed by the addition of ordinates with a fixed interval, and so eliminate variations of a definite period[1].

Thus if one of the component variations be

$$a_n \sin \frac{2\pi}{n} t,$$

where n is the period in years, t the time in years from a node, adding the ordinates with an interval of m years gives for this component in the secondary curve the sum,

$$a_n \sin \frac{2\pi}{n} t + a_n \sin \frac{2\pi}{n} (t + m),$$

and therefore a resultant of the same period,

$$2a_n \cos \frac{\pi m}{n} \sin \frac{2\pi}{n} \left(t + \frac{m}{2}\right) \qquad \ldots\ldots(3).$$

The amplitude is zero if $n = 2m$, and thus, in a curve of complex periodicity, the component $n\,(2m)$ is eliminated by the addition of ordinates with intervals of m years, while the amplitudes of all the other components are altered by the factor $2 \cos \pi m/n$; the phase of each is put forward by $m/2$ years.

This process has been applied to the curve of wheat-yield by taking the mean of *consecutive ordinates* whereby the variation of two years' period is

---

[1] I owe this method of dealing with complex periodic variations to Professor G. Chrystal of Edinburgh, who uses it for the discussion of the complex oscillations of the water of lakes under the name of the method of residuation.

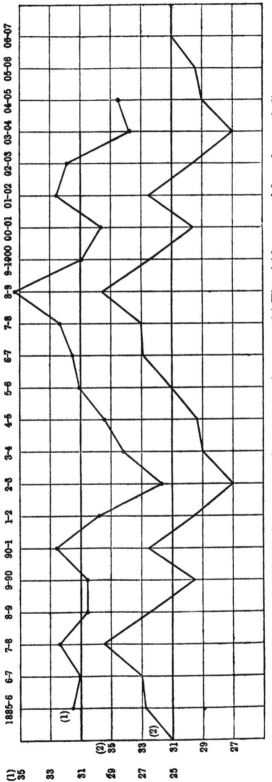

Fig. 83. (1) The actual yields of wheat: means of two consecutive years. (2) The yield computed from four periodic terms.

eliminated; the variations of other periods are put forward in phase by half a year, and the amplitudes are all altered as indicated by the coefficient. The result, which is represented in the upper curve (1) of Fig. 83, shows a very considerable simplification of the original curve of Fig. 82.

Further procedure by this process is not practicable for two reasons. First, we have only a single point for any year and no means of satisfactory interpolation. We can, therefore, only eliminate components of an even number of years by addition of ordinates. Secondly, if there be any accidental errors in the curve they get distributed over the sum in such a manner as to add to the complexity. There is a considerable reduction of the variation of ordinates upon taking the means of the ordinates (1) + (6) and also of (1) + (7), which points to a component of about 10 or 12 years' period. Taking this variation I have endeavoured to guess the components of curve (1), Fig. 83, by an inspection of the curve itself.

I have obtained in this way the periodic curve from which the ordinates of curve (2) are taken and of which the equation is:

$$y = 31 + 2 \cdot 8 \sin \frac{2\pi}{11} n + 0 \cdot 4 \sin \frac{4\pi}{11} n - 1 \cdot 2 \sin \frac{6\pi}{11} n + 1 \cdot 2 \sin \frac{8\pi}{11} n,$$

where n is the number of years (positive or negative) from the node at 1895–6. The successive ordinates correspond with the substitution of $n = 1, 2 \ldots$ in this equation.

The agreement between the curve constructed from this formula (curve 2) and the original (curve 1) representing two years' means of actual yield is remarkable. For the first four points the calculated yields certainly exceed the observed, and by an amount which reaches 3 bushels in 1887–8, but from 1889–90 to 1904–5 the mean value of the deviation between observed and calculated values, without regard to sign, is only ·67 bushel, and the maximum deviation + 2·3 bushels for 1902–3. The mean difference between the two curves is 0·1 bushel per acre. A shift of the curve through the space representing ·1 bushel would, therefore, slightly improve the correspondence. The representation of the real curve by the calculated one is, however, already as complete as one can expect any calculation of this sort to be, and pending the compilation of further data we may consider the average yield of wheat for successive two years to be effectively represented by a simple periodic variation of 11 years, with subsidiary harmonic components of 11/2, 11/3, and 11/4 years.

The original article in the Hann volume continued as follows:

"It is evident that if this representation is accurate the average yield for years to come should fall in with the same formula; in other words the formula allows us to predict the yield of future years. The prediction for 1905 on this basis is a yield that will give with 25·2, the yield for 1904, a mean for the two years of 29 bushels. In other words, the yield for the Eastern Counties for 1905 should be 32·8 bushels. The rainfall calculation for the same year gives 37·6 bushels. There is no doubt that the yield is large, but the Official

Report has not yet been issued, so the precise figures cannot be given. If 37·6 proves to be the correct figure there will be rather a large error (2·6 bushels) in the calculated harmonic curve."

The actual figure for 1905 turns out to be 32·0, so that it agrees better with the continuation of the composite curve than with the calculation from the autumn rainfall.

Since the article for the Hann volume was completed, I have extended the computation so as to represent the sequence of the yields for individual years by the sum of a series of six harmonic components of the 11 years' period. This extension of the reasoning accounts for the two year oscillation (which was eliminated from Fig. 83 by taking the average of consecutive years) and regards it as a "beat" between the oscillations of the fifth and sixth components of the 11 years' period. According to this analysis, of which the details are given in a paper read before the Royal Society on 17 May, 1906, the sequence of individual yields for the 21 years, 1885–1905, is represented with remarkable fidelity as the sum of six harmonic components of the 11 years' period, each with a node at the epoch 1895–6. The amplitudes of the several components are as follows:

Amplitude [31] + 2·9 + 0·5 − 1·8 + 2·8 + 1 + 1 bushel per acre.
Period ...   ...   11 $\frac{11}{2}$ $\frac{11}{3}$ $\frac{11}{4}$ $\frac{11}{5}$ $\frac{11}{6}$ years.
The − sign indicates a descending node.

Values derived from the addition of these six components are given in the second column of the table on p. 179, so that they may be compared with the actual values given in the third column. It will be seen that the direction of change from year to year is correctly given in every case and the agreement between the computed and the actual values is very close, except for the years 1887 and 1888, when the actual yield is too small and 1903 when it is too large. The way in which the minor fluctuations in the yield are indicated in the computed curve is very remarkable when it is considered that the components all refer to an 11 years' period or its submultiples, and the comparison extends over 21 years and not merely 11 years.

If the representation of the actual values by the computed curve have any real counterpart in the law of sequence of the yield of wheat it follows that the yield should repeat itself after 11 years. This is a result which is easily tested. In arranging the table on p. 179, I have put down not the actual yields but the excess or defect, + or − , from 30·8 bushels, and have grouped together the pairs of years with 11 years' interval; thus 1885 and 1896 are together, 1886 and 1897, and so on. The comparison between any year and the one 11 years later can thus be easily made. For three of the pairs of years out of the ten represented in the table the yield is repeated within a tenth of a bushel, for six it is repeated within a bushel. In every one of the ten pairs the yields are either both in excess or both in defect. The three cases where there is a substantial difference between the yields for a pair of years 11 years apart are those which contain the years 1887, 1888 and 1903 already mentioned.

That this simple principle of repetition after 11 years, which is easily traceable in the curves of Fig. 82, should not have been noticed until after a somewhat lengthy discussion is due principally to the fact that the yields for 1887 and 1888 are both substantially smaller than the computed values, and the curve, which should have had a salient point then, as it has in 1898 and 1899, has no conspicuous prominence, although its shape is not dissimilar in detail from the corresponding part 11 years later. It is curious, however, that in both the years named the yield, although too small from the point of view just mentioned, exceeded that computed from the previous autumn rainfall, so that the explanation of the defect, if indeed there be one discoverable, is rather recondite. The point is interesting because we have again in the current year (1906) a case in which the indication from the autumn rainfall is in the opposite direction to that from the principle of repetition after 11 years, and we do not yet know how it will be resolved[1]. It raises the question of the real meaning of the relation between the autumn rainfall and the wheat-yield of the subsequent year.

A paper by Mr E. Mawley (*Quarterly Journal Roy. Met. Soc.*, vol. XXIV, p. 75, 1898) "On Weather Influences on Farm and Garden Crops" suggested to me that a small autumn rainfall might be associated with a small summer rainfall following, and might therefore become an index of general meteorological conditions rather than itself a determining cause of the amount of the future crop. To illustrate this point I have included in the table on p. 179 in the same line with the figures for the crop for each year the excess or defect from the average of 20 years of the rainfall of each season, from the autumn before to the summer of the gathering of the crop. I have also added as an index of temperature-conditions the "accumulated temperature above 42° F." for the same seasons. The figures are taken from a paper on "The Seasons in the British Isles" in the *Journal of the Royal Statistical Society*, vol. LXVIII, part II (June, 1905).

Then at the foot of the table, following Mr Mawley's plan, I have given the average results for the five *best* wheat years and the five *worst*. The result is remarkable. Not only is the autumn rainfall deficient for the good years and excessive for the bad, but, with the exception of the accumulated temperature for the summer, which is above the average in both cases, the character of the season for the good years is the opposite of the corresponding season for the bad years in every case. Judging by the average results for the five years a good wheat yield is preceded by a dry and warm autumn, a rather dry and warm winter, a rather wet and cold spring, and a dry and slightly warm summer, while a bad wheat year has, on the other hand, a wet autumn of average temperature, a wet and cold winter, a dry and warm spring and a moist warm summer.

This bears out the suggestion that the autumn rainfall is in a way the key

---

[1] NOTE. December, 1906. The official estimates for the several counties are not yet published. The general opinion is that the yield was large but irregular as regards the Eastern Counties. [The rainfall won, see p. 180.]

to the subsequent seasons, but it still remains to consider whether these results are merely average results for the five selected years or whether they apply to individual years. Looking down the figures for the individual years it is clear from the juxtaposition of the signs that a wet autumn usually means a deficient crop and *vice versâ*; but it is also clear that a wet autumn is usually associated with a relatively dry spring and *vice versâ*, as is indicated in the averages for the five years. On seventeen occasions out of the twenty-one a deficient autumn rainfall has been followed by excess of rainfall in the spring or *vice versâ*. On two other occasions the spring rainfall has been normal or within a tenth of an inch of it, leaving only two exceptional occasions, forming the pair 1886 and 1897, when the heavy autumn rainfall was succeeded by spring rainfall above the average. On the other hand, on sixteen occasions out of the twenty-one the deviation of the winter rainfall from the normal has been in the same direction as that of the autumn rainfall. There appears to be no numerical relation between the amounts of excess or defect in either case, but it does appear that the suggestion of the five year averages that a dry autumn is followed by at least a rather dry winter and a rather wet spring and *vice versâ* is borne out in the large majority of cases.

As regards summer rainfall or accumulated temperature, however, there is little information to be got from the figures. Good wheat years are as a rule preceded by warm winters, and bad years by cold ones, but the figures are very irregular. Such a pair of years as 1888, 1899, as a glance at the figures will show, seems to make any generalisation about the relation of the temperature to the wheat or the other elements hopeless from the first.

There are, however, sufficient indications of underlying connections to encourage further investigation. The 11 years' periodicity in the yield, with its consequence, the repetition of values after 11 years, seems to be the most directly applicable of them. It would of course be unwise to regard the law of sequence thus indicated as definitely established.

The data available give us only approximately two 11 year periods, so that we are not able to apply an adequate test to the suggestion that the curve is in reality a curve of 11 years' period with its harmonic components. Many circumstances besides those associated with the practice of farming might be found to interfere with its persistence. It is possible that the concurrence in a node at the epoch 1895–6 is not mathematically strict; it is possible that the components of about four years', three years', or two years' period are not, strictly speaking, harmonics of the 11 year period. They may disclose increasing divergencies as the years accumulate. All that we can say is that for the 21 years for which data exist the representation of the values by a curve of 11 years' period is singularly close to reality, especially for the years 1889 to 1905. Further data, or perhaps data for some other quantity than wheat, may disclose a more definite result. In the meantime it is worthy of note that on the average for the seven counties comprised within England

East, the good wheat years have occurred with the following intervals from 1885[1] (itself a good year)

2 years,	1 year,	2 years,	1 year,	3 years,	2 years
1887*	1888*	1890	1891*	1894	1896*

completing 11 years, and then again

2 years,	1 year,	2 years,	1 year,	3 years
1898*	1899*	1901	1902*	1905

As this order of succession has now occurred twice, it seems worth while to consider the question in relation to the rotation of crops:

TABLE

Year	Yield above or below 30·8 bushels per acre computed from six harmonic components of eleven years' period	Actual Yield above or below 30·8 bushels per acre	Rainfall in inches above or below the average of the 20 years 1881–1900				Accumulated Temperature above 42° F. in day degrees above or below the average of 20 years, 1881–1900			
			Autumn	Winter	Spring	Summer	Autumn	Winter	Spring	Summer
1885	+4·5	+3·0	−0·3	+1·3	+1·0	−3·2	+25	+46	−104	−89
1896	+4·5	+3·6	−0·9	−1·9	−0·2	−1·4	+113	−4	+33	+63
1886	−0·7	−1·6	+5·0	−0·6	+0·6	−0·9	−146	−86	−49	−60
1897	−0·7	−2·3	+1·9	+2·5	+0·1	−0·9	−72	−25	+14	+110
1887	+4·8	+2·1	−0·7	+0·3	+0·2	−3·0	+221	−49	−116	+163
1898	+4·8	+5·5	−2·5	−1·2	+0·5	−1·6	−58	+71	−68	+2
1888	+4·7	+1·1	−0·9	−1·7	+0·4	+3·4	−244	−81	−87	−213
1899	+4·7	+3·4	−2·4	−0·4	+0·7	−2·8	+244	+139	−21	+201
1889	−2·0	−1·5	−2·2	−1·6	+2·0	+1·8	−40	−46	+67	−25
1900	−2·0	−3·3	+0·2	+3·0	−1·2	+0·2	+99	−36	−102	+167
1890	+0·2	+1·0	−0·7	−0·7	0·0	+1·5	−68	−1	+5	−119
1901	+0·2	+1·0	−2·8	−0·4	+0·2	−2·7	+112	+43	−45	+71
1891	+2·4	+2·7	−2·0	−2·6	+1·0	+1·3	+122	−9	−143	−96
1902	+2·4	+2·6	−2·8	+0·4	+0·4	+0·9	+79	+8	−51	−121
1892	−4·3	−4·8	+0·6	−0·2	−0·3	+1·9	+56	−20	+35	−147
1903	−4·3	−0·4	−3·3	−0·8	+0·5	+5·0	+37	+128	+34	−85
1893	−4·4	−5·6	+2·6	+0·6	−3·7	−0·6	−153	−28	+260	+179
1904	−4·4	−5·6	+1·4	+0·7	−1·0	−1·6	+19	−76	+26	+91
1894	+1·1	+0·3	−0·7	−0·5	+0·8	+1·1	−28	+44	+36	−67
1905	+1·1	+1·2	−3·8	−1·4	0·0	−1·2	−6	−11	+19	+187
1895	−4·1	−3·1	+0·3	+0·4	−0·7	+1·4	−27	−69	+62	+15
1906	−4·1	?	−1·6	+1·2	?	?	−127	−20	?	?
Five best wheat years		+3·6	−2·2	−0·8	+0·6	−1·5	+89	+49	−61	+16
Five worst wheat years		−2·5	+1·0	+0·9	−1·4	+0·3	−1	−46	+56	+85

[1] The years marked * are those which according to the computed curve give yields exceeding the average by at least 2 bushels; for the others the theoretical yield is not much above the average.

YIELD OF WHEAT

APPENDIX

Table of Yields of Wheat for England East compared with Yields computed from Autumn Rainfall and from the 11 Years' period extended to 1921. An asterisk (*) means that the Computed Yield differs from the Actual Yield by as much as 2 Bushels per acre.

	Rainfall Spring	Rainfall Previous Autumn	Wheat from Rainfall	Wheat Actual	Wheat from Period		Rainfall Spring	Rainfall Previous Autumn	Wheat from Rainfall	Wheat Actual	Wheat from Period
1885	+1·0	-0·3	29·9*	33·8	35·3	1907	+1·1	+1·2	27·3*	35·3	35·3
1886	+0·6	+5·0	18·2*	29·2	30·1	1908	+1·2	-1·7	33·9	33·4	30·1*
1887	+0·2	-0·7	30·8*	32·9	35·6*	1909	+0·9	-2·3	36·1*	33·8	35·6
1888	+0·4	-0·9	31·3	31·9	35·5*	1910	+1·0	+1·1	28·4	28·4	35·5*
1889	+2·0	-2·2	34·2*	29·3	28·8	1911	-0·4	+0·9	28·9*	31·5	28·8*
1890	0·0	-0·7	30·8	31·8	31·0	1912	-1·0	+0·8	28·9	30·7	31·0
1891	+1·0	-2·0	33·7	33·5	33·2	1913	+0·9	-0·1	31·1	31·3	33·2
1892	-0·3	+0·6	27·9	26·0	26·5	1914	+0·6	+0·4	29·9*	33·0	26·5*
1893	-3·7	+2·6	23·6	25·2	26·4	1915	-6	-1·5	34·1*	31·2	26·4*
1894	+0·8	-0·7	30·9	31·1	31·9	1916	+50	-20	32·3*	26·6	31·9*
1895	-0·7	+0·3	28·6	27·7	26·7	1917	+2	-1	31·0	28·1	26·7
1896	-0·2	-0·9	31·2*	34·4	35·3	1918	+12	-10	31·7	32·2	35·3*
1897	+0·1	+1·9	25·1*	28·5	30·1	1919	+4	+45	27·0	27·5	30·1*
1898	+0·5	-2·5	34·7	36·3	35·6	1920	+17	-57	35·8*	31·7	35·6*
1899	+0·7	-2·4	35·1	34·2	35·5	1921	-47	-75	37·5	36·2	35·5
1900	-1·2	+0·2	28·8	27·5	28·8	1922		-61	36·3		28·8
1901	+0·2	-2·8	35·4*	31·8	31·0						
1902	+0·4	-2·8	35·5*	33·4	33·2						
1903	+0·5	-3·3	36·6*	30·4	26·5*						
1904	-1·0	+1·4	26·3	25·2	26·4						
1905	0·0	-3·8	37·6*	32·0	31·9						
1906	-1·0	-1·6	32·8	33·5	26·7*						

On the revision of normals for the district it is probable that ·4 has to be added to Autumn Rainfall differences before 1905 and for 1906 and 1907, not for 1908 and following years.

The differences from normal are given in inches up to the autumn rainfall of 1914. Beginning with the spring rainfall of 1915 the differences are given in millimetres. The normal rainfall for Autumn is given as 6·91 in. in the Weekly Weather Report, 1914, no. 47.

13. METEOROLOGY AND AGRICULTURE

A LECTURE DELIVERED IN CAMBRIDGE ON 23 JULY, 1912,
AT THE CONFERENCE OF AGRICULTURAL TEACHERS

THE IMPORTANCE OF METEOROLOGY FOR AGRICULTURE

THERE is a widespread opinion, more often hinted than expressed, that meteorology is of great importance in agriculture; yet when one looks around for monumental evidence of its achievements in this country, except for some successes in forecasting, there is little to be seen. I want you this afternoon to turn your attention to that opinion, and ask you in what way meteorology is or can be of importance to agriculture. The opinion is probably based upon the undeniable fact that the weather is of the greatest importance to the growth of crops. Remember that meteorology is not the weather, but the study of weather. The inference is a natural one that the study of weather must be a matter of practical importance to those who are interested in the growth of crops; yet most people when suddenly challenged about the weather in relation to crops would call to mind, not the orderly sequence that makes a crop almost a certainty, but the losses due to adverse weather. They would recollect what they pay for insurance against hail, the havoc of spring frosts, the diseases of prolonged wet weather, the lack of fodder in a dry season, the well-grown crops that could not be harvested, the destruction caused by gales and snowstorms, and (when it comes to transporting perishable goods) by fogs. The obvious effects of all these destructive agencies are intensely real; they generally seem to be quite wilful and unnecessary, and from time immemorial they have produced a lasting impression upon the human mind, not really so deep-seated as that of the orderly sequence which keeps hope immortal in the farmer's breast, but much more easily expressed in words when the subject is mentioned.

THE LOSS DUE TO TEMPESTUOUS WEATHER

The destructive aspect of the relation of weather to crops is of undeniable importance. It has to do with what I may call the tempests, or intemperance of the weather, the bad tempers of a climate, the incidents which, as a rule, take us unawares when we are unable to protect ourselves.

In a newspaper the other day I noticed that according to the census of production the gross value of the agricultural produce of Great Britain in

1908 (excluding holdings of less than an acre) was £150,800,000 divided into the following categories:

		£
Farm crops		46,600,000
Fruit, flowers and timber ...		5,200,000
Animals		61,400,000
Wool		2,600,000
Dairy produce		30,000,000
Poultry		5,000,000
		£150,800,000

I have often wondered what addition could be made to a figure of this kind if we could rely upon the weather being always benignant, but I have not yet found the means or the opportunity of compiling the necessary information. By way of making a guess, let us say that the difference of value between a good crop and a bad crop is 25 per cent., and that, taking one year with another, half the crops are bad and half good. We might in that way arrive at $12\frac{1}{2}$ per cent. as the deduction that is generally made on account of the tempestuousness or intemperance of the weather. This amounts to about £20,000,000 a year. The first view that I wish to bring to your notice is that this estimate is probably not excessive and that in dealing with the losses due to "tempests" of all kinds very large sums are involved. I should be greatly obliged if some one of my hearers would go through the returns and find out the gross value of the best harvest and of the worst harvest of each separate crop in the past 20 years, and add together all the highest values to form a "highest possible," so that the deficiency in any year from that standard might be computed, and so give us an estimate of the actual losses due to tempestuous weather.

Please note that for convenience I use the word "tempest" for any exaggeration of intensity that mars the beneficence of weather, by the malevolence of destructive energy or by the paralysis of necessary activities.

Meteorology in Relation to Tempests

What then has meteorology, the organised study of weather, to say to agriculture about these tempests of wind, rain, snow, hail, lightning, frost, fog or drought? First of all, it keeps a record of the circumstances of their occurrence in the shape of daily maps for different regions of the globe and notes the frequency and distribution of tempests of different kinds, so that every one may know at least what difficulties they have to overcome or to yield to. Secondly, it recognises that all these tempestuous events are examples on the large scale of processes that can be described in the language of physics and represented by experiments in the physical laboratory, and it therefore seeks an explanation of all these exaggerations by diagnosing the physical processes which they follow and the dynamical causes which produce them. To that end it keeps continuous records of a number of the physical properties of the air at the surface and makes frequent soundings of the upper air. The purpose of this endeavour is that knowing how the tempests are produced we may be able to say with greater confidence when they will occur. Thirdly, it uses its knowledge to forewarn the public of the approach of the different kinds of tempestuous weather by storm-warnings for gales and by daily forecasts of other kinds of weather. It is this division of meteorology which is concerned with the daily weather service of the Meteorological Office, the economical basis of which is that it is useful for the public to be forewarned of coming weather.

FORECASTING

I shall not spend much time over this side of the subject. It will in time make its own appeal to the general public. I will show only a succession of four maps of the United States showing the invasion of the Southern States of the Union by a cold wave. About these maps we may fearlessly assert that when the front of the depression has been ascertained the rest of it follows and it obviously is an advantage to know what the succession is going to be and to take all possible precautions against the damage which is inevitable if plants and animals are left exposed. It is a fair illustration of the possibilities of the daily forecast.

I have drawn the illustration from maps of the United States because the "tempests" of cold are peculiarly violent there; we have similar occurrences in our own country, though generally on a less violent scale.

The proper method of dealing with this part of the subject is for the agriculturist to keep himself informed of the sequence of events as represented in the Daily Weather Report, and to acquaint himself with the principles upon which the official forecasts are compiled. He will then be in a position to make full use of a telegraphic forecast. It must be remembered that in forecasting, with few exceptions such as rain after a period of drought or fine weather for harvests, "tempests" are the matters of real importance.

Consequently the daily forecast is useful in so far as it enables the farmer to anticipate the approach of a tempest of wind, rain, hail, snow, lightning, frost or fog[1]. Success in its application requires continuous attention to the changes going forward and hence occasional forecasts are for the most part ineffective. It is satisfactory therefore to be able to record that the use of the Daily Weather Map is becoming more and more extended as a branch of "Nature Study" in schools, and that a knowledge of this aspect of meteorology is gradually becoming widespread.

In illustration of the method which we employ I have obtained some copies of the Daily Weather Report of to-day, giving the results of observations in

[1] Upon this claim Mr T. H. Middleton, of the Board of Agriculture, remarks that it is too modest and adds, "These are doubtless the important things, but from February 1 to May 31, when the crops are going in, an intelligent use of forecasts relating to normal weather should assist the farmer in tilling the land to the best advantage."

Western and Northern Europe and the Atlantic Islands at 7 a.m. this morning which have been charted, discussed and printed at the Meteorological Office by 11.40 a.m. and are here for distribution at 2.30 p.m.

The Clemency of Weather

That completes what I have to say for the present about the £20,000,000 worth of damage done to crops by inclement weather. I want next to say something about the £150,000,000 worth which we owe to its clemency. How does it come about that our climate, against which at times so much is said, is good enough to grow crops to the value of about £3. 3s. per acre taken over all, and why £3. 3s., why not £3. 10s. or £4? The limitation of output is due partly to soil, partly to climate, and partly as I have said to tempests. The agriculturist has to take account, if possible, of all three, at any rate of the first two. It is the business of meteorology to give such an account of the material which the regular study of the weather affords that an agriculturist may think out whether, apart from any modification of the soil, his output has reached the limit which the climate permits. This is a part of the subject which has been frequently commended in word, but is yet very little attended to in deed. Perhaps it would have been better if I had referred to his predecessor's output, for it will certainly be urged that the long experience of a practical farmer of ordinary intelligence enables him to adjust his practice to the average sequence of weather without the necessity for putting down any of the figures on which meteorology is based. While this cannot be denied, it is fair to say that it is a pity that the benefit of experience should be lost for want of putting it on record. The application of meteorology to agriculture might have as an alternative title "What the farmer knows and won't tell." To enable the long experience of the past to be used for the profit of the present is the true object of the application of scientific method. For that purpose meteorologists all over the country have been encouraged, now for some 60 years, to make collections of statistics, chiefly expressed in the form of monthly summaries, of pressure, temperature, humidity, and so on, which are fully represented and illustrated in our Monthly Weather Report; and with the same object in view the Meteorological Office has supplemented its daily task of guarding as far as possible against the tempests of weather, by preparing week by week a body of homogeneous statistics of weather (for each of 12 districts in the British Isles), and thus exhibiting in its Weekly Weather Report the course of the weather in the most effective manner for comparison with the experience of the agriculturist.

After these prolonged efforts, extending over 62 and 34 years respectively, we are still rather vague about the importance of meteorology to Agriculture. We are proposing to hold a special meeting of the International Meteorological Committee next year to consider the subject, at the suggestion of the International Institute for Agriculture of Rome; but when I ask what use our practical farmers make of the information which we compile with so great labour, I get in reply some reference to forecasting with which I sympathise, but practically nothing about statistics. People, in fact, remember without difficulty the losses which adverse weather has inflicted; the records of its benign influence are not thought of. As of other benefactors, it may be said of the passing years:

"The evil that they do lives after them,
The good is oft interred with their bones."

Therefore, the second view that I wish to bring to your notice is that it is time for the bones of meteorological statistics, dry as they may appear, to become a fertiliser like phosphates or nitrates and in their own way to help towards adding a shilling per acre to the average earnings of British land, or say £2,000,000 to the annual value of its total produce.

THE LACK OF ORGANISED KNOWLEDGE

What is wanted in the first instance is to get our knowledge properly organised. Everybody knows that to bring a crop successfully to maturity we require, (1) time, (2) sunlight, (3) warmth, (4) moisture in the atmosphere (humidity), (5) moisture in the ground (rainfall), but very little is actually known at present as to how much time? how much sunlight? how much warmth? how much humidity? or how much rainfall? When these questions can be answered, the expression of the relation between the amount of the crop and the amount of the various elements will probably be simplified.

The third view that I wish to offer to you is that our ignorance of these matters is really surprising. Only a few days ago I received a letter from the editor of a Dublin newspaper asking which of the three years '75, '76, '77, and which of the three '79, '80, '81, was the best for agriculture in Ireland, and in reply I could only ask another question, "What, so far as weather is concerned, is a *good year* for agriculture?" If I knew what was a "good one" I might hazard a guess at which was the best. Really this is a question of the proper organisation of statistics, upon which Mr Yule will speak to you. I will put before you diagrams representing first the average year (Figs. 84 to 87, pp. 192 and 193), and secondly, quarter by quarter, the variations from the average of the weather conditions as regards sunshine, temperature and rainfall for the various districts of the British Isles since 1878 (Figs. 88 to 93, pp. 194–199). It ought not to be difficult to set against each year a note giving its character as an agricultural year, but it has never been done, and I do not know where to find the information in a compendious form. Clearly some further means of discrimination is necessary.

THE SEPARATE ELEMENTS OF A GOOD YEAR

I have already said that a good crop depends upon time, sunlight, warmth, humidity and rainfall. Let us consider some points with reference to the several questions taken separately.

1. *How much time is required for a good crop?*

One end of the time is obviously when the good crop is gathered, the question is when ought one to count from?

From time to time I read in the newspapers accounts of how the crops are getting on and I find generally a tendency to credit the plants with too short memories. Taking a diagram of the relation of the wheat crop in the Eastern Counties to the rainfall of the previous autumn it is easy to see that the wheat never forgets and hardly ever forgives the treatment which it has received in the autumn. And another diagram of successive crops for the 21 years 1885 to 1905 seems to show that the wheat actually remembers what its crop was 11 years ago and tries to imitate it.

From my own casual observation of the behaviour of fruit trees, I suggest for your consideration that apple trees, pear trees and plum trees have a very vivid recollection of what the weather was last year[1], and that the yield of this year will depend more on last year's fine summer than on this year's weather. Of course, an act of violence, like a late frost, will do a great deal of damage in any case, but how far a fine summer is numerically represented by a good crop next year is a question which is worth your consideration.

2. *How much sunlight is required for a good crop?*

To this question I can give no answer. Mr H. T. Brown's experiments on the assimilation of carbon by leaves lead one to conclude that full sunlight is much more powerful than is necessary for the ordinary processes, and that a good deal of the plant's skill and intelligence is devoted, not to using, but to protecting itself against the excessive power of direct sunlight.

When this question comes to be discussed we shall want to learn not only what total duration of sunlight is necessary, but what the distribution of sunlight should be during the plant's life, in order that the best result may be obtained. Of what doses should the total amount be made up? And the same is true of the other meteorological elements.

3. *How much warmth is necessary for a good crop?*

Here is another very wide field that has been inadequately explored. The aggregate warmth of any locality is a very complicated matter. A diagram representing the distribution of maximum and minimum temperature round the parallel of 50° N. lat. shows climates that are very different, yet nearly all will grow wheat.

On the other hand a corresponding section west to east across the British Isles shows very little difference in different parts, yet the differentiation of crops is most noteworthy.

Many years ago it was pointed out by De Candolle that below certain

[1] Mr Middleton attributes a remarkable wealth of roses in this year, 1912, to last year's fine summer, and Mr Cave, of Ditcham Park, has called my attention to the remarkable load of fruit this year on the beech trees in his plantations. It seems probable that a heavy crop of hips and haws is a reminiscence of the summer of the preceding year rather than a prognostic of a severe winter to come.

temperatures plants cannot grow and that warmth must be reckoned as "accumulated temperature." At the instance of Sir John Lawes and Sir Henry Gilbert of Rothamsted the Meteorological Office expressed its statistics of warmth as accumulated temperature. As to temperature, let me show that although a certain amount of warmth is indispensable, it is possible to have too much of a good thing. A great aggregate of warmth in the spring and summer in the Eastern Counties means a barley crop deficient in quantity—perhaps as regards quality it may be otherwise.

Let me also refer to a remarkable result put forward by Mr Unstead in the *Geographical Journal*. He finds that the total warmth, or accumulated temperature, required to mature a wheat crop depends on the duration of daylight in the period of growth as well as the mean temperature, being less when daylight is prolonged. His figures lead him to the conclusion that the accumulated temperature necessary for a crop in any locality is proportional to an empirical index-number made up by adding to the mean temperature during growth the average number of hours of darkness in the locality for the same period.

4. *How much humidity is necessary for a good crop?*

Here I think we want the assistance of agriculturists and horticulturists before we can attempt an answer. I can show you the humidity at different times of the day in different months of the year at Kew, Falmouth, Aberdeen, Valencia. Note how great is the difference at different times of the day. No one can doubt that humidity is of the greatest importance to plants; but is it the humidity of midday or of midnight that must be reckoned?

5. *How much rainfall is necessary for a good crop?*

I suppose that for any crop the distribution of rainfall in the year is of more importance than the total amount. A large part of the Canadian middle West can grow wheat with a rainfall of about 10 inches a year because the rain falls in the summer months when the wheat requires it. The former wheat crops of Cyprus were connected with the winter incidence of rainfall.

THE PRACTICAL BEARING OF THE SPECIFICATION OF A GOOD YEAR

From what I have said you will understand that we have still a good deal to settle before we can say what is a good weather-year for agriculture, but with some perseverance it could be done. What then? Would it be of any use? I have discussed that aspect of the subject too with various persons, and will conclude by giving you the views at which I have arrived.

THE FIXITY OF THE FARMER'S SYSTEM

First, I argued the question with a prominent agriculturist on the basis of the long memory of some plants and asked whether it would not be desirable in any particular year to lay stress on the crops that were not already subject to some ascertained disability: for example, when the autumn has turned out wet, to substitute some other crop for wheat. Weather goes more or less in cycles although they are not easily defined, and as regards any particular field it must sometimes happen that it grows wheat in a good wheat year, roots in a good root year, beans in a good bean year, and grass in a good grass year; whereas another field, with the ill luck which pursues some people, will grow wheat in the bad wheat year and so on with the other crops in rotation. On the four years the second field will be say 25 per cent. worse off, as compared with the first, which happened to take its rotation in the proper years.

At the end of the discussion I found that the course which the farmer had to follow was so regulated that, without regarding the weather, he had practically as many simultaneous equations to satisfy as he had quantities at his disposal. He had to take his rotations in a certain order, to keep a certain balance between his crops in order to feed his stock and he had little left to his discretion—he was bound to go through with it and chance the weather. So I waited, and in the meantime did what I could in the direction of forecasting, which is at present the only generally recognised point of application of weather knowledge.

The Possibilities of the Use of Statistics

Part of our programme in that division of the work is to prepare a special series of forecasts for the harvest season. The forecasts are as a rule for 24 hours, though in certain circumstances they run for two or three days or even for a spell of weather. Through the Board of Agriculture a circular inviting applications for these forecasts reached a farmer in the north, who was good enough to give me his view upon the question. I am always glad to know what people really think of these attempts to save something of the £20,000,000 of annual loss. The view expressed by this commentator was that forecasts for a day or two, even if they were generally accurate, were of no practical use to him, but if I could have told him beforehand that we were going to have no summer in 1910 and a perfect summer in 1911, or that April and May of 1912 were going to be very dry, he could have turned the information to good effect. I thought then that my opportunity had come and that I was going to learn what freedom of discretion there was in agriculture to which meteorology could appeal; so, with a cheerful anticipation, I wrote and asked my correspondent what he would have done, different from what he did, if I had told him beforehand what he allowed would have been really useful. You will recollect that I had previously been cornered by the view that the course of agriculture did not permit of change on account of the weather, and here was a possibility of getting out of the corner.

But I grieve to say I have had no reply and the question still remains unanswered. I am very sorry for it because for this exploration there is, on a very moderate estimate, prize-money to the extent of £20,000,000 a year.

If my correspondent had shown me how the country could have saved £5,000,000 by knowing what crops were going to fail or succeed on account of the wet summer of 1910 or the dry summer of 1911, it would have been my business to let the public know that there is such a prize to be won, and to ask for the co-operation of the educational world in an endeavour to win it. It may require, it probably does require, co-operative organisation for the world; and that means the co-ordination of our knowledge of the meteorology of the land, and a large extension of our knowledge of the meteorology of the sea, especially of the polar regions. But if the enterprise required is world-wide, the prize is also for all countries and must be reckoned in the aggregate not in tens but in hundreds of millions. And besides I could have said to my candid agricultural friend, "if you see such great utility in the knowledge which we are hoping to get, but do not now possess, will you not help in making use of that which we already have? If I cannot tell you that your hay crop will be spoiled, can you make no use of the information, that you might have had in November 1910, that the wheat crop was safe?"

Mr R. H. Hooker in a paper before the Royal Statistical Society has shown how in the East of England, the yields of the several crops are dependent upon temperature, rainfall and sunshine in different parts of the year, or more, before the harvest. From the correlation coefficients which are there given and which are represented in Figs. 95 and 96 many useful suggestions may be derived. Certainly another step is possible, you can at least help us to find out what is a "good weather-year" for any given crop in any district. The information that I have put before you suggests that we may have to take into account information that is already available long before the crop is actually harvested —sometimes before it is sown. And if by laying our heads together we can turn that information to good practical effect by choosing suitable lands for cropping in a special way in different years we shall not only get the encouragement to go on with the attempt to solve the more difficult problem of forecasting the seasons, but we shall fairly earn the money that such an enterprise requires.

SYLLABUS OF QUESTIONS ON THE RELATION BETWEEN METEOROLOGICAL INFORMATION AND AGRICULTURAL PRACTICE

By way of suggesting questions of practical interest in agriculture which require meteorological statistics for their answer, I have drawn up a syllabus, which is reprinted here, for use in opening a discussion at the Meeting of the British Association in Dundee. It need not be supposed that any single meteorologist or agriculturist is in a position to give answers to the questions for any specified district at the present time. The intention of the syllabus is rather to indicate the lines upon which the information could in time be organised. Nor is it contended that a single central office could deal adequately with the information from all districts of the country. The situation indicated is the development of local organisations for separate districts.

Climate in the Average Year

1. What crops are grown in what counties? and why? What limits the selection of crops in the different counties?

2. What aspects or positions are specially favourable for crops of different characters? fruit? vegetables? trees?

3. What is the relation between climate and crops in respect of (1) quantity, (2) quality, (3) time of harvest?

4. How much of (1) time, (2) sunlight, (3) warmth, (4) humidity of the air, (5) rainfall, is required for the best crop of any kind: for the best forest growth?

5. What is accumulated temperature, and what is its use in agriculture? Why is the temperature of 42° F. selected as the base for calculating accumulated temperature? Is the same base suitable for all crops?

The Variation of Seasons

6. What effects do the differences of season as represented by rainfall, moisture, air-temperature, earth-temperature, snow, sunshine, have upon the crops in respect of (1) quantity, (2) time of harvest?

7. Is the effect different according to the nature of the soil?

8. What deviations from normal weather represent (1) a good year for any particular crop? (2) a bad year?

9. Can treatment of the soil protect against the effect of adverse seasons?

10. Can any economy be secured by adjusting the courses according to a knowledge of the weather of past seasons and their known influence upon the various crops?

11. What action should be taken in the case of ground being (1) under snow, (2) under water, (3) dry, for an exceptionally long period?

12. What amount of rain can fall within a day, two days, three days, a week, a fortnight, a month, without causing floods? in winter? after a frost? in summer?

13. What is the local relation between rainfall and water supply in wells and rivers?

14. What is the longest drought that can be supported without suspension of farming operations of various kinds (growing crops or stock) on chalk or porous subsoils? on clay or non-porous subsoils?

15. How long can young growth do without water before hope of recovery has to be given up?

16. What is the relation between the prevalent diseases of plants and animals and the previous weather?

Inclement or Tempestuous Weather

17. What is the actual and percentage annual loss of crops on account of inclement weather?

18. What are the facts and figures upon which the rates of premium for insurance against damage by rain, hail, snow, frost, lightning, fogs, and drought are based?

19. What are the earliest dates in autumn and the latest dates in spring when killing frosts have been experienced? How do they depend upon situation?

20. What precautions other than insurance can be taken against the loss of crops or stock?

DIAGRAMS OF AVERAGE ACCUMULATED TEMPERATURE, SUN-
SHINE AND RAINFALL FOR FOUR SECTIONS OF THE BRITISH
ISLES; AND OF QUARTERLY VALUES OF MEAN TEMPERATURE,
RAINFALL AND SUNSHINE FOR THE TWELVE DISTRICTS FOR EACH
OF THE YEARS 1878–1910 REFERRED TO ON PAGE 185

The following diagrams represent in graphical form statistics given in
Appendix I of the Weekly Weather Report for 1910 published by the
Meteorological Office.

For the tabulation of statistics in the Weekly Weather Report the British
Isles are divided into twelve districts 0–11 enumerated in the Figs. 88–93.
Each district comprises several stations—for instance Scotland East in 1910
comprised Gordon Castle, Nairn, Aberdeen, Balmoral, Crieff, Leith, West
Linton and Marchmont; the mean for the district is the mean of all these
stations. Similarly for the other districts. (See any issue of the Weekly
Weather Report.)

Districts 1–5 are combined to give a general mean for "Eastern Districts,"
and 6–10 to give a general mean for "Western Districts."

Figs. 84–87 give in graphical form the averages of "accumulated temperature," sunshine and rainfall for 25 years for four divisions of the country,
namely: Scotland N. (Northern Section), the Channel Isles (Southern Section), the combination of Scotland E., England N.E., England E., Midland
counties and England S.E. (Eastern Section), and the combination of Scotland W., England N.W., England S.W., Ireland N. and Ireland S. (Western
Section). They show also by dotted lines the weekly values and hence the
variations from the average of the 25 years, of temperature, sunshine and
rainfall for the year 1912 from the beginning of March to the end of the
41st week. "Accumulated temperature" is thus explained in the Weekly
Weather Report:

The tables of accumulated temperature are designed to give persons engaged in
agriculture better means of estimating the manner in which vegetation is affected
by temperature than that afforded by the more usual methods of treating the readings
of the thermometer. They show the combined amount and duration of the excess
or defect of the temperature above or below a suitably fixed standard or *base
temperature*. For *Districts* cumulative values for the whole period since the commencement of the season and the year are published in addition to the values for
each week. The base temperature adopted is 42° F. as being nearly equivalent to
6° C. This has been considered by continental writers on these subjects to be the
critical value, the temperature above which is mainly effectual in starting and maintaining the growth, and in completing the ripening, of agricultural crops in a
European climate. This base is also convenient as being 10° F. above the freezing
point on the Fahrenheit scale.

The accumulated temperature is expressed "Day-degrees"—a Day-degree
signifying 1° F. of excess or defect of temperature above or below 42° F. continued
for 24 hours, or any other number of degrees for an inversely proportional number
of hours.

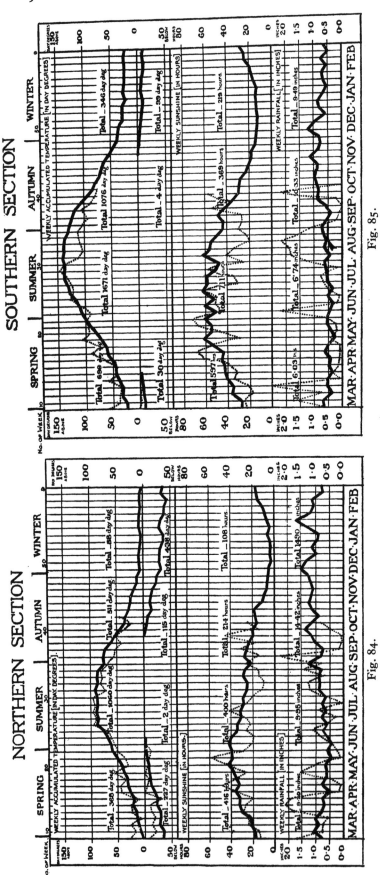

Fig. 84.

Fig. 85.

COURSE OF THE SEASONS IN THE BRITISH ISLES. WEEKLY AVERAGES, 25 YEARS 1881–1905.

Fig. 84. The heavy lines indicate the averages for the 25 years, and the dotted lines the values for 1912 from the beginning of March to the 12th October.
There is a deficiency of temperature from the 30th week onwards with deficiency of sunshine throughout the period. Heavy rainfall in April and at the end of August followed by conspicuous dry weather in September and early October.

Fig. 85. The deficiency of temperature from the 30th week and the general deficiency of sunshine extends to this section with heavy rain in March, a wet August and again a dry September followed by heavy rain in this section with heavy rain in early October.

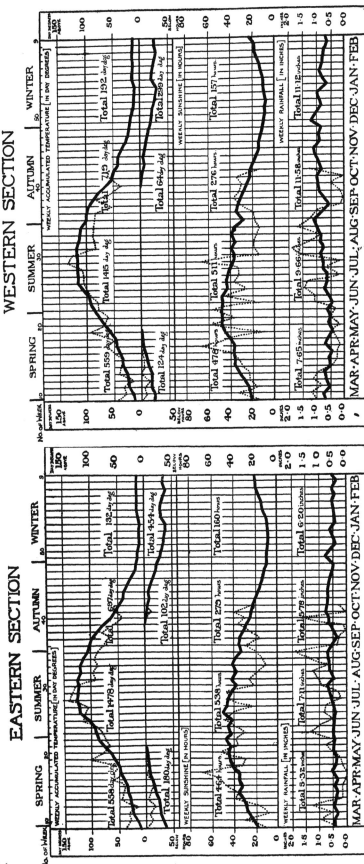

Fig. 86.

Fig. 87.

COURSE OF THE SEASONS IN THE BRITISH ISLES. WEEKLY AVERAGES, 25 YEARS 1881–1905

Fig. 86. There is a serious deficiency of temperature from the 30th week onwards with again general deficiency of sunshine and a dry September.
Fig. 87. The deficiency of temperature in the Western Section after the 30th week completes this deficiency for the whole country together with the deficiency of sunshine. A dry September preceded by some intervals of heavy rainfall is again shown.

MEAN TEMPERATURE OF THE BRITISH ISLES, 1878–1910

Differences from the normal of the values of the successive quarters

Districts 0–5

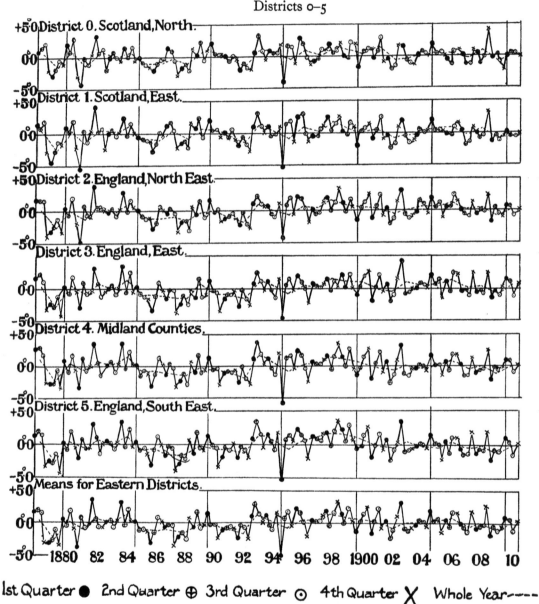

1st Quarter ● 2nd Quarter ⊕ 3rd Quarter ⊙ 4th Quarter ✗ Whole Year - - - -

Fig. 88. Temperature variations are shown by differences from the average
in degrees Fahrenheit.

The dotted line refers to the mean for the whole year. Thus in Scotland East 1898 was the
warmest whole year of the series with a mean temperature 1°·6 above the average and 1879
was the coldest year with a mean of 2°·6 below the average.

Of individual quarters in Scotland East the warmest first quarter was January—March 1882
which was 4°·2 above the average, and the coldest was January—March 1881 which was
5°·4 below the average.

MEAN TEMPERATURE OF THE BRITISH ISLES, 1878–1910

Differences from the normal of the values of the successive quarters

Districts 6–11

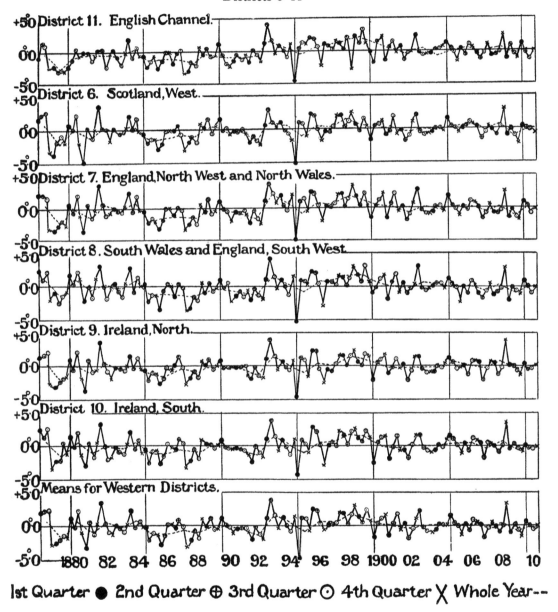

1st Quarter ● 2nd Quarter ⊕ 3rd Quarter ☉ 4th Quarter ✗ Whole Year --

Fig. 89. Temperature variations. A continuation of Fig. 88. Districts 6–11.

Thus in Scotland West as in Scotland East the warmest year was 1898 (+1°·7) and the coldest 1879 (−2°·1). In the same district the warmest first quarter was in 1882 (+3°·7) and the coldest in 1895 (−5°·0).

RAINFALL IN THE BRITISH ISLES, 1878–1910

The Rainfall of the successive quarters expressed as percentage of the normal

Districts 0–5

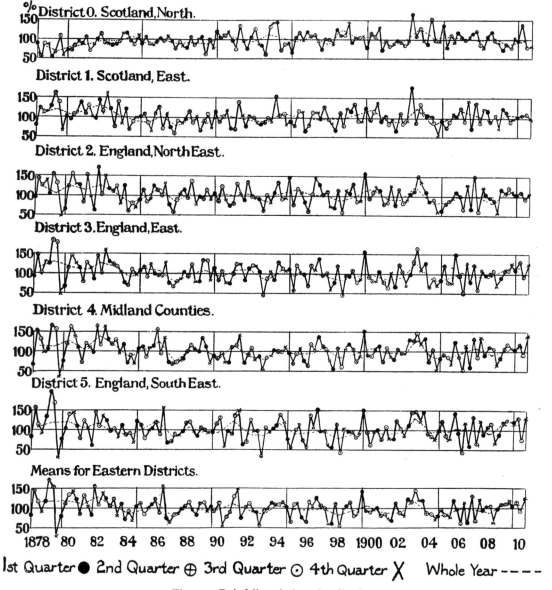

1st Quarter ● 2nd Quarter ⊕ 3rd Quarter ☉ 4th Quarter ✗ Whole Year – – – –

Fig. 90. Rainfall variations for districts 0–5.

The differences from the average are shown by expressing the values for each year and quarter as percentages of the average. In Scotland East 1882 was the wettest year with 131 per cent. of the average rainfall and 1887 was the driest with 79 per cent. of the average. In the same district the first quarter in 1903 had 175 per cent. of the average rainfall and the fourth quarter of 1904 only 52 per cent.

RAINFALL IN THE BRITISH ISLES, 1878–1910

The Rainfall of the successive quarters expressed as percentage of the normal

Districts 6–11

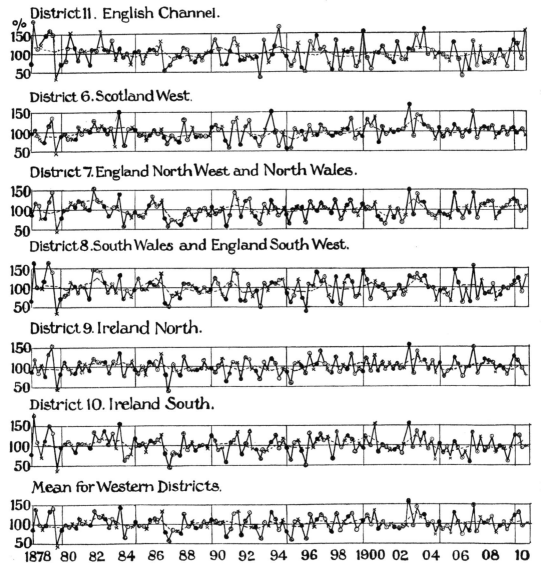

1st Quarter ● 2nd Quarter ⊕ 3rd Quarter ⊙ 4th Quarter ✕ Whole Year - - -

Fig. 91. Rainfall variations. A continuation of Fig. 90 for districts 6–11.

In Scotland West the wettest year was 1903 with 134 per cent. and the driest was 1895 with 80 per cent. of the average. In the same district the most rainfall in the first quarter of 1903 (172 per cent.) and the least in the fourth quarter of 1879 (45 per cent.).

SUNSHINE IN THE BRITISH ISLES, 1881–1910

The number of hours of sunshine in successive quarters expressed as percentage of the normal

Districts 0–5

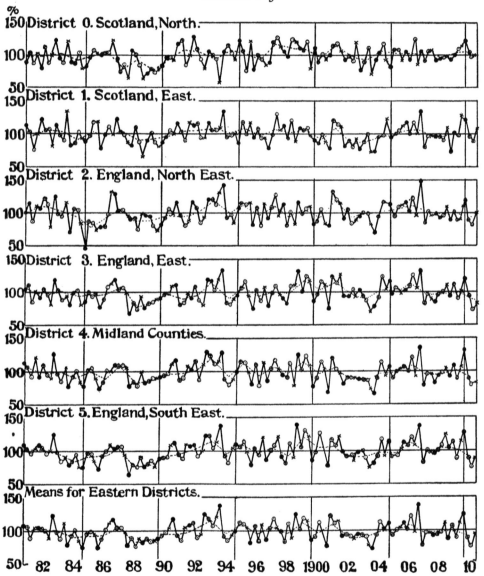

1st Quarter ● 2nd Quarter ⊕ 3rd Quarter ⊙ 4th Quarter ✗ Whole Year --

Fig. 92. Sunshine variations for districts 0–5.

The differences from the average of the mean sunshine values are represented as percentages of the averages. Thus in Scotland East the yearly values ranged from 110 per cent. in 1901 to 85 per cent. in 1902. In England South East in 1899 the percentage was 124. The sunniest third quarter, July—September, occurred in districts 0, 1 and 2 in 1897, in 3 in 1906, in 4 in 1906 and 1899, and in 5 in 1899. The least sunny third quarter occurred in 1888 in districts 2, 3, 4 and 5.

SUNSHINE IN THE BRITISH ISLES, 1881–1910

The number of hours of sunshine in successive quarters expressed as percentage of the normal

Districts 6–11

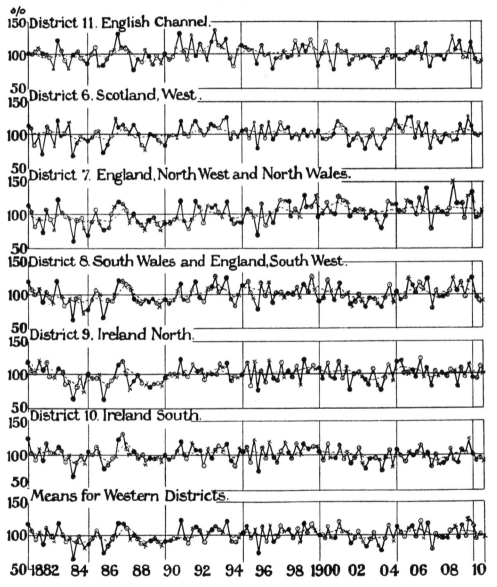

Fig. 93. Sunshine variations. A continuation of Fig. 92 for districts 6–11.

In Scotland West the yearly values ranged from 113 per cent. in 1893 to 89 per cent. in 1884. The sunniest third quarter occurred in districts 6, 7 and 9 in 1906, in 8 in 1899, in 10 in 1898 and 1900, whilst in district 11 the third quarter of each of the years 1898, 1899 and 1900 has 120 per cent. of the average amount of bright sunshine.

When the temperature during any period remains either wholly above or below the base temperature, the difference between the base temperature and the mean temperature of the day gives the correct accumulated temperature. In other cases this difference gives only an approximate value for the accumulated temperature, not departing greatly from the truth—the deviation depending on the greater or less extent of the variations of the temperature above or below the base. Further, since the mean between the maximum and minimum of any day is nearly equal to the mean temperature of the day, the difference of the mean of the maximum and minimum from the base will also give, directly, an approximation to the accumulated temperature for the day.

The present series of the Weekly Weather Report dates from 1878 and the Appendix referred to gives for each district the mean temperature and mean rainfall for each quarter and for each whole year from 1878 to 1910 with the differences of these means from the averages for the 25 years 1881–1905. For sunshine the comparisons begin in 1881. Figs. 88–93 give a graphical representation of these differences.

SEASONS AND CROPS—ENGLAND EAST
Explanation of Figs. 94—96

Mr R. H. Hooker has calculated the correlation coefficient between the abnormalities of weather and the corresponding deviations from the average of certain crops:

A correlation coefficient is a number representing the relation between the deviations from the normal of corresponding values of two elements.

When the correlation coefficient is 0 the deviations have no relation one to another.

When the correlation coefficient is 1 the deviations are strictly proportional the one to the other.

When it is − 1 they are proportional but opposite in sign.

When it is greater than ·3 there is definite probability of a connection between the deviations.

When it is greater than ·5 the connection is certain.

In order to calculate the coefficients the means of weekly values of rainfall and accumulated temperature have been taken for groups of eight weeks. The first group is from the ninth to the sixteenth week of the year preceding the crop year, the next from the thirteenth to the twentieth, the third from the seventeenth to the twenty-fourth and so on, and the correlation is taken between the deviations of each group from its average for the 21 years 1885 to 1905, and the deviations of the crop from its average for the same years.

The accompanying diagrams exhibit (1) the weekly averages (25 years) of the elements from the ninth week of spring until the 44th week of the following year (Fig. 94) and (2) the values for the correlation coefficients found by Mr Hooker (Figs. 95 and 96). Certain general conclusions are drawn from these, and are given below.

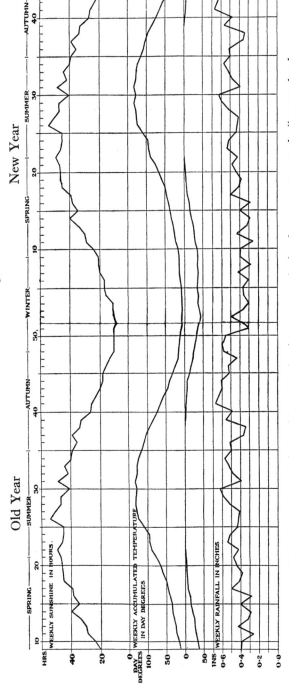

Fig. 94. Normal Weekly Duration of Sunshine, Weekly Accumulated Temperature and Weekly Rainfall for District 3, England East

Note: The second curve under weekly accumulated temperature in day-degrees represents the "accumulated temperature below 42°" and an aggregate of cold.

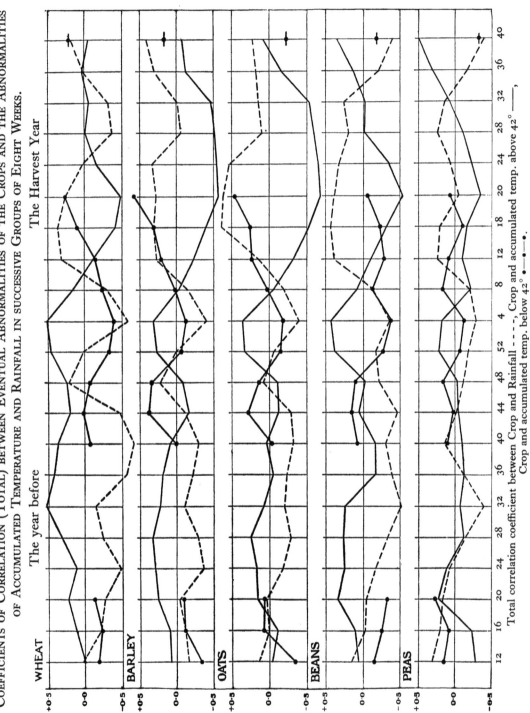

COEFFICIENTS OF CORRELATION (TOTAL) BETWEEN EVENTUAL ABNORMALITIES OF THE CROPS AND THE ABNORMALITIES OF ACCUMULATED TEMPERATURE AND RAINFALL IN SUCCESSIVE GROUPS OF EIGHT WEEKS.

The Harvest Year

The year before

WHEAT

BARLEY

OATS

BEANS

PEAS

Total correlation coefficient between Crop and Rainfall - - -, Crop and accumulated temp. above 42° ——, Crop and accumulated temp. below 42° •——•.

Fig. 95.

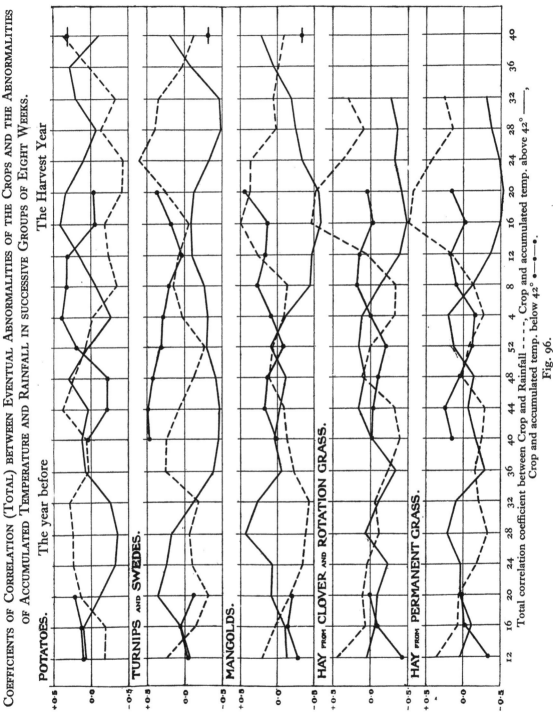

Fig. 96.

In referring to the curves, the following technical terms are used:

1. A "crucial" period is a period when the coefficient of correlation between the crop and one or other of the elements of weather is greater than 0·5.

2. Warmth is "deficient" when the temperature is below the figure for the corresponding week of the average year and is "unusual" if it is above the average.

3. Rainfall is "heavy" if it is above the figure for the corresponding week of the average year, and is "light" if it is less than the average.

For *Wheat* the weather should be unusually warm and dry except during the greater part of the spring of the crop year, when it should be cold and wet. Warmth is specially important in the previous mid-winter and mid-summer.

Dryness is specially important in the previous autumn and winter. During the harvest summer temperature is of no account but dryness is favourable.

A dry autumn is crucial, as are likewise unusual warmth in the previous summer and winter and deficient warmth in the late spring of the crop year.

Barley. The most important consideration is "deficient warmth" in the spring and summer of the crop year associated with more than average rainfall, except in the middle of summer. Unusual warmth and dryness of the preceding year are favourable.

The autumn and winter are of little importance, but a spell of cold in the late autumn is good, so is a spell of dryness in mid-winter.

The deficiency of warmth in spring and summer of the crop year is the only crucial consideration.

Oats. Still more markedly than barley, oats are favoured by "deficient warmth" and "heavy rainfall" in spring and summer of the crop year. The previous spring and summer are of very little importance but dryness is favourable and warmth in a less degree. The autumn and winter are also of little consequence, but "unusual warmth" in mid-winter is on the favourable side.

The only crucial considerations are "heavy rainfall" and "deficient warmth" in late spring and early summer of the crop year.

Beans. Rainfall below the average during the previous spring, summer, autumn and winter is good, followed by "heavy rainfall" in spring and summer of the crop year, which is associated with deficient warmth in the late spring. Otherwise temperature is of little consequence, except in mid-winter when unusual warmth is favourable.

The crucial periods are in the dryness of the previous summer and autumn, and the deficiency of warmth at the end of spring.

Peas appear not to be dependent upon abnormalities in the year.

Potatoes are exceptional in being favoured by cold and wet in the summer and autumn of the previous year. They like a cold winter followed by dryness in the late winter, spring and summer, with warmth in the spring.

No period is, however, crucial.

Turnips and *Swedes* are favoured by deficient warmth continuously from the mid-summer of the previous year till the crop is gathered. Rainfall is of little consequence until the summer of the crop year, when there is a crucial period of heavy rainfall and deficient warmth.

Mangolds, like so many other crops, are favoured by a warm dry summer in the previous year, but a crucial period is the middle of the spring of the crop year when deficient warmth and heavy rainfall are indicated from the beginning of spring until the crop is gathered.

Of all crops Mangolds are the least dependent upon autumn and winter weather.

Hay. The chief characteristic about hay is the striking advantage of wet, cool weather in the spring of the harvest year, which culminates in a crucial period of heavy rainfall in the middle of spring. The previous year is hardly a matter of interest for the hay from seeds. *Permanent pasture* appears to be favoured by relative lightness of rainfall in the summer and autumn of the previous year, but the indication is not very strong.

Mr Hooker has followed up his investigation by examining the correlations when the years subsequent to 1905 are taken into account. Reference may be made to his Presidential Addresses to the Royal Meteorological Society in 1921 and 1922.

14. THE WEATHER

A Lecture Delivered at the Princess May Road Junior Commercial Institute, Stoke Newington, London, N., 6 October, 1913

When I was invited by the Education Officer of the London County Council to take part in the inauguration of the New Scheme of Institutes I was at a loss to know what to say and indeed what to talk about. For the last 13 years my life has been devoted to organising the study of the weather on sea and land in the interest of the general public. The official name for the study of the weather is meteorology and the centre for the organisation of the study of the weather in the interest of the public is called the Meteorological Office. These are the long hard names which we use, I suppose, in order to maintain our relations with science on the one side and with Government on the other. We are the Meteorological Office, not the Weather Office, because it is by the dry light of the science of meteorology and of that alone that we regard the weather; we are the Meteorological Office, not the Meteorological Institution, because we are created and maintained by Government. Not many of the British public were aware of that until I told you: all the same, it is the fact.

But what has meteorology to do with evening institutes where grown boys and girls can continue for their own advantage and their country's the education of which they have received a good deal and perhaps retained a little? It would not be unreasonable, since the Government regards the study of the weather as of sufficient public importance to justify the maintenance of an office at the public expense and has so regarded it for 60 years, that some opportunity should be provided for making the people acquainted with what it is all about. Whatever your walk in life may be, remember that you will have the weather always with you, in fact you will never be so nearly independent of the weather as you have been while you were at school. It is one of the peculiarities of the British Islands that on about 99 days out of 100 the weather does not matter. If you will think, the number of days that are remembered in British history because of their weather is very few; there are certain occasions which we treasure in our memories of fierce wind or intense cold or deluges of rain, but a long lifetime does not collect many. In this metropolis the independence of the weather is made more nearly complete by good roads, good drainage, good means of locomotion (which are, however, apt to be crowded on a wet day)—I wonder if I ought to add good boots, good overcoats and umbrellas. And then, when you are at school good building that keeps out the weather, good heating that keeps out the cold, and good lighting whenever you want it, have enabled you to snap your fingers at the weather.

If I had set out to find the group of people who are likely to think they need not trouble themselves about the weather, I should ultimately find my way to an evening continuation school of the County Council of London— and here I am, not apparently for that reason but for the very opposite.

Do you remember Shakespeare's suggestion of a fine starlight night?

> How sweet the moonlight sleeps upon this bank.
> Here will we sit and let the sounds of music
> Creep in our ears;
> Sit, Jessica. Look! how the floor of heaven
> Is thick inlaid with patines of bright gold.
> There's not the smallest orb, which thou behold'st
> But in his motion like an angel sings,
> Still quiring to the young-eyed cherubims;
> Such harmony is in immortal souls.
> But, whilst this muddy vesture of decay
> Doth grossly close it in, we cannot hear it.

That might perhaps have been written on the bank of the Thames, but the occasions must be quite exceptional when this passage recalls the experience of Stoke Newington, N.

So independent of the weather are the Englishmen, the Scotchmen, the Irishmen and our fellow subjects in the dominions beyond the seas who manage our education for us, that I think I am not saying anything but what is true when I sum up the result of their management thus: the higher the education the less the study of the weather. I believe I represent in my own person the whole of the separate provision for University teaching in meteorology in the British Empire. I am the one subject of His Majesty who has a University appointment in the study of the weather, and out of respect for my country, I won't mention the salary.

In the schools it is certainly different and for a very good reason. The average grown-up Englishman, to judge by his daily paper, takes a great interest in the weather. I know of no daily newspaper that does not find room for about half a column of information about it, most of which comes from my office, though it is sometimes translated into journalese by " Our own meteorologist." It is, in fact, the most unvarying of all news items in the newspapers, excelling in that respect even the stock-markets, the turf, cricket and football, for with the weather is no annual holiday, no Sunday, no Christmas Day, and no Good Friday, and when His Majesty proclaimed two bank-holidays for the celebration of his coronation he bestowed two days of real hard work upon his faithful Meteorological Office.

It is a little remarkable that the British public is so vigorously insistent upon having the latest information about the weather and that the Government should be willing to supply it for them, and yet that so little trouble should be taken to enable the public to understand what it all means.

If the subjects of an evening institute are to be limited to vocational or scholarship subjects there is not much to be said for meteorology. For the staff of my Office and a few others, about a hundred people I suppose all told, in these islands it has become vocational, but that is all. In the Universities non-vocational studies have little chance, whatever public interest there may be in them, so you must not expect scholarships in meteorology. There ought to be a considerable number of openings with the Press, but a pressman never regards any subject as outside his vocation; he knows what the public wants and what it appreciates and is always equal to an emergency. If the subject and he do not understand one another, the subject has to give way.

Occasionally I have the advantage of seeing what items they select out of the reports that I write for the Government or the lectures that I give in public, and on these occasions, let me confess, I never fail to wish that there were more encouragement for the study of the weather, even in continuation schools. For if it is not a vocational subject the weather sooner or later obtrudes itself into everybody's vocation and everybody wants to know about it. Look where you will, in literature, in art, in the Navy, in the Army, in local Government, in medicine, in law, in commerce, in industry, in agriculture, in transport, in the field, in the mine, on the mountain, in the valley, in the east and in the west, in the north and in the south, on land, on sea and in the air, in work or in play, in war or in peace, everywhere and in all things you will find the weather refuses to be disregarded. Even in the Civil Service:

'Neath the baleful star of Sirius
When the postmen slowlier jog,
And the ox becomes delirious,
And the muzzle decks the dog.

You can take no part in life without being affected by the vicissitudes of weather. It has affected literature in all ages. It has been the theme, not only of newspapers of to-day, but of the poetry of all time.

I am quoting something familiar to most schoolboys and schoolgirls when I recite the lines:

Sweet and low, sweet and low,
Wind of the western sea,
Low, low, breathe and blow,
Wind of the western sea.
Over the rolling waters go,
Come from the dying moon and blow,
Blow him again to me,
While my little one, while my pretty one sleeps.

This is a place of examination, and I will pay my respects to the local deity: Will you kindly tell me why the poet spoke of the "wind of the western sea" and not of the eastern sea? Whoever gives the right answer will get some marks for the study of weather.

So I will assume that you, with me and the whole world, have some interest in the study of weather and I will go on to explain what the organised study of weather as represented by the Meteorological Office means. From the point of view of the general public it is divided into two great sections which are not really independent though for the purposes of study we may treat them as being so.

There is first the preservation of an accurate and trustworthy memory of the events of weather year in and year out and in every part of the world. Those of you who go into business will soon realise the immense advantage of an accurate memory; the one indispensable person in an office is the one who always knows where things are and remembers what has been done. If you are to preserve an accurate memory you must keep a careful record of events and so far as the weather is concerned it is the business of the Meteorological Office to keep that record for all parts of the British Isles and of the High Seas. Other countries keep records in a similar way for their own lands. We put our records into print and exchange them; so now, since every civilised country has its weather office, every country is able to profit by every other country's memory as well as its own.

I shall show you one example of long memory, namely, the record of the rain that has fallen over London in the last 100 years. You may know that the rainfall is measured by catching what falls within a measured area, an eight-inch or a five-inch circle, and seeing what depth it comes to, so we give the rainfall in inches. Fig. 97 represents the fall in inches for each month from 1 January, 1813 to 31 December, 1912, and a very interesting and instructive record it is. It is also rather perplexing in its irregularities, but that is another story. In the same way we collect from many stations similar measures of rainfall, temperature, wind and other elements of weather.

I shall show you another example of meteorological memory of a very detailed character (Fig. 98). It shows the rainfall at Kew Observatory for the last 10 years in a form which is designed to answer the question "How often does it rain at different times of the day and night?" How often has it been necessary to put up an umbrella between 6 and 7? 7 and 8? 8 and 9? 9 and 10? and so on throughout the day? Now let me remind you that everyone of you ought to be able to answer that question for some of the hours because you have been going to school or doing something else at certain hours and you have known, once upon a time, whether it rained or not—probably you have forgotten, and if you want to refresh your memory here is the means of doing so. The figures give the number of times that rain has fallen in the hours of the day indicated by the numbers in the top line for each month of the year. There is a separate line of figures for each month. By grouping the figures into regions by lines which we call isopleths, we can get a good idea as to which are the hours most likely to be rainy and which most likely to be fine. September comes out remarkably dry, March and October on the whole, wet. The columns are added up and show in the bottom line the number of times it has rained, on the average, at a definite hour throughout the year. From these figures we may conclude that the best hour to go to business is that which centres round 10 o'clock (the normal hour for the Civil Service) because it is on the average the driest hour of the day; and the best time to go home at night, if you want to keep "dry," is 1 o'clock in

The rainfall in inches for each month is shown by the length of a column as measured against the scale of inches at the side, and the fall in inches for each year is given in figures above the heads of the columns

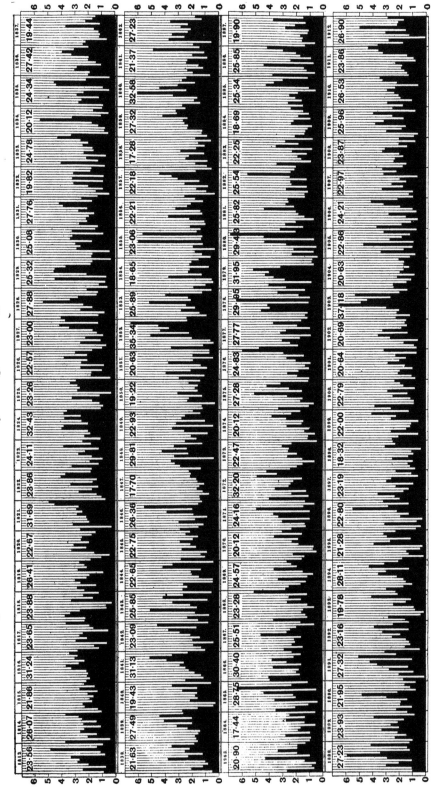

Fig. 97. One hundred years of rain for London; fall for each month 1813–1912

	1	2	3	4	5	6	7	8	9	10	11	12	13	14	15	16	17	18	19	20	21	22	23	24	Monthly totals
January	19	27	27	19	23	20	20	22	22	21	17	20	16	21	17	18	26	21	26	24	23	25	22	22	518
February	16	16	19	20	18	23	22	22	20	19	19	21	18	20	18	31	29	25	18	18	20	16	15	18	481
March	34	32	27	23	23	33	35	28	25	19	23	31	35	35	39	38	37	41	35	27	33	33	33	30	746
April	17	21	22	25	19	19	25	26	21	15	18	19	28	28	32	38	30	24	25	21	23	22	18	19	555
May	21	18	21	26	28	22	19	24	24	18	26	25	23	23	22	31	22	24	23	28	18	29	24	26	575
June	26	25	31	28	31	29	33	23	28	22	25	27	32	32	27	32	32	28	30	25	21	23	23	27	660
July	18	17	15	28	23	15	15	15	11	16	17	24	24	20	25	31	24	25	25	28	25	22	21	27	489
August	18	24	32	31	23	22	18	18	22	19	13	25	25	27	30	28	31	27	27	18	20	22	19	22	561
September	11	14	16	20	16	17	20	20	16	16	14	20	18	15	20	15	20	21	20	16	14	14	11	14	398
October	27	37	36	39	42	42	36	32	35	31	33	36	29	28	28	37	38	37	36	34	35	22	31	22	803
November	31	27	27	28	28	24	19	21	24	20	22	31	29	24	31	27	24	26	26	20	27	26	28	28	618
December	28	29	33	34	31	26	23	31	28	24	21	25	28	28	32	34	38	33	31	25	39	36	28	27	712
Hourly totals	266	287	306	308	305	292	285	282	276	240	248	304	305	301	321	360	351	332	322	297	292	290	273	273	7116

Fig. 98. Frequency of hours which gave rain at Richmond in ten years.

the morning. Perhaps that is why the closing hour for public houses is 12 h. 30 m. a.m.

The other chief function of the Meteorological Office is to forecast the weather of the future and it is this which attracts the most attention from the public. I shall only explain quite briefly what we do in this respect. We have about 30 observers at points distributed over the three kingdoms who, at 7 o'clock each morning, read the barometer and thermometer, note the wind and the state of the sky, and at 7.15 send a telegram to the Office reporting these observations which are exclusively figures. I cannot help thinking that some people are under the impression that what we do is to have a weather-wise person at each of these outlying stations to telegraph us a "tip" as to what he thinks the weather is going to be, but that is not so. We do not ask them their opinion as to what the weather is going to be, we form our own by putting on a map the observations which they send us and seeing how the pressures, temperatures, etc., are grouped. By international agreement we exchange our observations with those of other countries and thus we get a map of the present weather extending from the middle of the Atlantic to the confines of Russia, and from Spitsbergen in the North to Madeira in the South.

The map is prepared about 9.30 a.m. and forecasts are drawn from the study of it. A complete report is ready by 11 a.m., and goes to the lithographer in the basement of the Office at South Kensington who returns us 850 copies by half-past twelve. Before one o'clock more than half of them are in envelopes and messengers have started to catch the 12.30 post at the G.P.O. so that the information may reach as many as possible on the same day. A similar process is gone through in connection with observations at 1 o'clock p.m. and 6 o'clock p.m. except that there is no issue of a lithographed map, the results are communicated to the newspapers.

These two departments, the meteorological memory and the daily map of the weather, are both well adapted for the educational study of weather. The latter is now being used for that purpose in a large number of secondary schools in all parts of the country.

While the two functions already referred to define the immediate objects of the Office from the point of view of the general public, there is another point of view, namely, that concerned with the science of meteorology. It comes into office work as part of the responsibility for endeavouring to explain the succession of events which are enshrined in the Office memory and is intended to lead up to the extension of forecasting beyond the day or few days from which at present the forecasts are available, to the seasons, the year and so on. Why is it that one month is wet, another dry? Why should one August be dry and warm, a second wet and cold and a third dry and cold? Before we can give an answer we must understand what it is all about and find out the relations of events which are at present apparently unrelated. That is the problem of the science of meteorology. It involves first of all an organised meteorological memory for the whole earth because we are sure that the variations of weather in one part are closely related to those of another part and there is no part of the world that we can safely leave out—not even the poles. It also means a knowledge of the atmosphere at all levels and the reference of the events observed to the laws of mechanics and physics and to the calculations of mathematics. It is a slow and a difficult process which has been made much harder by some of our predecessors in the investigation guessing at facts to satisfy explanations instead of patiently extending the region of observations till they were sure of their facts.

I shall show you a few illustrations of the things which meteorologists record for their own information and of the subjects which are engaging their thoughts.

First the clouds—we all know how diversified they are and we want to find out why. We can do a little more than simply stare at them, we can photograph them; we can find out where they are, how high and how far off, how they are moving and also how they change. Then we try to make out why they change.

Next, we try to find out how the wind varies at high levels even when there are no clouds to be seen. For this we send up a small balloon filled with hydrogen and watch it with a telescope that enables us to make out a plan of its track and so tells us how the wind changes in the upper layers. I have the results of a pair of experiments of this kind represented by photographs of models (Figs. 99 and 100) which show by the length of card the wind at successive heights, one of them shows a gradually increasing southerly wind until a height of 6 miles is reached; then it falls off; the other shows an easterly wind which falls off rapidly and is replaced at about 2 miles by a westerly wind that goes on increasing as we rise.

My last slide is obtained by means of balloons which carry instruments for recording pressure and temperature and go to great heights, as much sometimes as 20 miles, far beyond what a man could endure. It is a photograph of a glass model of a block of air 15 miles high over the British Isles on 27 July, 1908 (Fig. 23). It has lines drawn on it which show what one would see if one could tell how cold the air was by looking at it; the broad band near the ground shows where the freezing point of water was found. Up to about 5 miles the colder layers of air are met with in succession until a remarkable change occurs, the cold layer over one part turns up on end over the other part so that we ultimately come to a new region where it gets no colder in the vertical but shows some change along the level.

Now let us go higher still to 100 or 200 miles; for this information we depend upon what mathematics can infer for us from what we can observe down here. We find first a mixture of gases and water vapour; then the water vapour stops, then the nitrogen and oxygen disappear, and finally we are left with hydrogen and helium alone (Fig. 67).

These facts are for the most part new and they open up a vista into the unknown which is at present the possession of only a few enthusiasts. On them we ponder day by day, week by week, year by year, while the ceaseless ebb and flow of the daily weather passes under our view. If we speak of them in public we are generally met with a rebuke for speaking in unintelligible language, and a demand for something which the man in the street can understand. It is an impossible prayer. You might as well ask the astronomer to

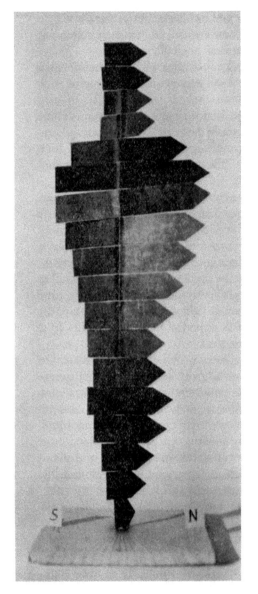

Fig. 99. Increase of S.-wind from the surface upwards on 1 Oct. 1908.

Fig. 100. Reversal of an easterly wind at four kilometres, 6 Nov. 1908.

The models show the direction and velocity of the winds for each kilometre of height by arrows which fly with the wind and represent its velocity by their length.

talk about the heavens without using such technical terms as the sun, or moon, planet, or comet. It is the wrong way to advance; names are always difficult so long as they are unfamiliar; the proper way is to regard the weather, which is a subject of universal interest, as a subject of universal instruction. If in time the subject should be taken up in a school like this I cannot say whether it will or will not add to your wages, but I can say that it will give you information which touches every walk of life; it will give you an insight into the mysteries of nature which are scrutinised in every country of the globe, and if you carry away nothing else you will read a deeper meaning in the words of the Hebrew psalmist which you can find written up in the Meteorological Office:

> The heavens declare the glory of God:
>   And the firmament sheweth his handywork.
> One day telleth another:
>   And one night certifieth another.
> There is neither speech nor language:
>   But their voices are heard among them.

15. THE ARTIFICIAL CONTROL OF WEATHER

Lecture before the Cambridge Aeronautical Society, 9 March, 1921

THE control of weather has been a subject of vivid interest from the dawn of history down to the present day. It is woven into the fabric of every form of civilisation. The claims of the rainmaker are modern; but they are not exclusively modern, and are therefore not to be regarded as one of the many signs of the progress of physical science in civilised nations. They are equally characteristic of some of the most primitive civilisations of Central Africa. Quite deep down in human nature is apparently the feeling that if man cannot himself control the weather at least he knows who or what can; and he can bring influence to bear upon the spirits of the air or upon the conditions which those spirits personify that will guide the control in the manner desired. It is perhaps only people with long experience in the study of weather who really feel that the appropriate attitude with regard to the vagaries of weather is not to control but to observe them, and try to understand them. The control of weather is engineering not meteorology. It is no more meteorology than the building of the Channel tunnel is geology.

Control according to ancient mythology

Few subjects of speculation are more interesting than the system of control that is indicated by the mythology of the Greeks. Zeus controlled the thunder, and his deputy Aeolus the winds. You might have to sacrifice a good deal to appease either of them but always, sooner or later, as everyone knows, they are appeased. I am sure that the Harpies with their rapacity and their dirt had charge of what we call "line squalls"; and, for reasons given elsewhere[1], I think it not unlikely that the Gorgon was the personification of the Mediterranean cyclone, which happens only in the winter and has a habit of being followed by a wind cold enough to turn everything into stone.

The position in the eighteenth century

The discovery of the laws of motion of the planets in the seventeenth century, with which the great name of Newton is associated, produced a profound effect upon the thought of the eighteenth. The literature of the period resounds with the conception of "laws which never shall be broken"; but the weather was regarded as still at the immediate pleasure of the Almighty Law-giver in Whom had become gathered all the several powers of the Greek immortals. The actual direction of operation might still be entrusted to an angel, a subordinate power, who

> Pleased the Almighty's orders to perform
> Rides in the whirlwind and directs the storm.

[1] "Illusions of the Upper Air." Royal Institution, 10 March, 1916.

The supposed effect of noise in the Middle Ages

The transition through the Middle Ages from the mythological position of the Greeks and Romans to the theistic position of the eighteenth century was very gradual, and is perhaps not altogether complete in some parts of Europe even at the present day.

In the often-quoted inscription of mediaeval bells "vivos voco, mortuos plango, fulgura frango" we find an indication of the power of sound over the more violent expressions of weather; and it is curious that when it is a question of controlling weather the loudness of the sound is regarded as being specially influential. I suppose that in the course of countless generations of men, the persistent experience of the association of rain with the noise of thunder has made it impossible for any one of us to take a really unbiassed view of the place of noise in the control of weather; and certainly we ought not to be intolerant of the man in the street, who is "not to be convinced by any arguments" that there is no connection between the two. It is not surprising that, when fire-arms were invented, the guns of the peasants who tended the vineyards of Europe took the place of the church-bells in the protection against damage by hail and were let off as a precaution. The spirits of the air were not forgotten. It became the practice to save from Candlemas the ends of candles wherewith to load the guns in order to have something more directly operative than sound in driving away the malevolent spirits that were supposed to be in control of the hail-clouds.

The consciousness in modern times of increased power of directing the energy of nature

In the present generation not only are the laws of motion of the heavenly bodies regarded as never to be broken, in spite of the fact that Einstein and others may alter the form; but there are many new laws of physics and chemistry which have an equal claim to be regarded as inexorable in the study of weather, among them particularly the law of conservation of energy: and moreover the powers of the laboratory and the workshop have become so much enlarged that the new spirit of humanity is not disposed to take the vagaries of the spirits of the air "lying down." If we really understand the forces of nature, we ought to be able to direct the operations; and we find a disposition to ask whether we cannot ourselves take over the forces of the air, and if not, why not?

It is not exactly the magnet's question:

> If I can wheedle
> A knife or a needle,
> Why not a silver churn?

but another question: "If I can do almost anything I please in the laboratory and the engineering workshop, what is there to stop me complying with any specification of conditions for the open air?" Many opinions of the futility of human effort have already been proved to be wrong. Kelvin is said to have proved that flying was impossible; yet now we fly. Kelvin did not know the possibilities of the internal combustion engine; and in like manner every other awkward corner in directing the operations of nature may be turned by new inventions.

The scientific mind

It has been drummed into us of late years that in order to maintain our place in the universe we must use science as our guide to action. Everybody must be educated in scientific method, and the scientific attitude must describe our position with regard to general policy. Those who, like myself, have been mixed up with science from early days at school accept this kind of appeal with instinctive approval, and a scientific society may be expected to do the same. I have often wondered what the attitude towards the experience of life would be with no scientific training, with no habit of considering any statement from the point of view of how it might be proved. Presumably, in that case, scientific fact ranks as the expression of an author's opinion, against which can be lawfully and decisively set the opinion of someone else of greater material or moral worth. A few days ago I happened to be turning over some old letters and found one of 1855 addressed to the Royal Society from a Government official of some distinction, himself a Fellow, protesting against the Society adopting the theory of the independence of pressure of water vapour and dry air, because in his opinion any such theory was erroneous. What right had he to an opinion? If a person in his position, a professed man of science, could waste time in denying one of the fundamental laws of physical science instead of proving it, what might not the ordinary layman deny? One can understand that there may be many people who are prepared to insist that the earth is flat and that Newton was quite wrong in his law of universal gravitation. Indeed, I have a friend who likes to think that it will be possible sooner or later to find a means of annulling weight and who is probably now regarding Einstein as the saviour of mankind from the thraldom of gravity. Reasoning is not always conclusive. Here is a specimen derived from the time when we used to work with box kites that is very interesting. By varying the shape of a kite you can make it work with its string at a steeper and steeper angle with correspondingly less pull on the wire; it is evident that, if you carry your experiments far enough, you can get the string vertical and have no pull on the wire at all. If you have any considerable experience with kites you will know that that does happen sometimes; then, of course, you can cut the wire and you will have obtained a kite that will fly without any string. [Q.E.F.]

OFFERS TO CONTROL THE WEATHER

In the course of my experiences at the Meteorological Office, I have had to be responsible for considered opinions on many offers to control the weather in some form or other.

This was specially the case, but not by any means exclusively so, during the war, when it was represented in the highest quarters that the course of events showed clearly that our enemies had learned to produce rain at their pleasure and it was our duty to go one better than they by adopting certain forms of apparatus of which the efficiency was said to be beyond doubt. There is, of course, a certain amount of danger on the one hand that with the habit of regarding the atmosphere as something of which one had to ascertain the laws by observation, experiment and reasoning, one might lose sight of the possibility of control; but, on the other hand, it was generally public money that was required to set the proposed machinery in motion; and to divert prematurely towards control money that was given for discovering or verifying laws would be to waste the money as well as to make a public exhibition of oneself.

It is astonishing what people will suggest. I have not yet recovered from the astonishment of receiving from the head of an important educational establishment an offer to send to the Office a member of his staff who had an infallible method of forecasting the weather. The suggested advantage of the arrangement was that the forecasters of the Meteorological Office, if they happened to be of military age, could leave their work, and "join up." The poignancy of the suggestion, so far as I was concerned, was that just before the war the copyright of a method, which I judged to be identical, was offered to me and I, wrapping myself in my scientific dignity, had not thought proper to buy it.

I propose to put before you some of the proposals for controlling the weather which came before me.

General objects of control

The objects of the various suggestions were curiously limited. I do not recollect any suggestions for beginning where Nature begins and turning winter into spring or summer for a particular district by warming the open air or the open sea or for drying the roads by operating on the humidity of the open air. The objects to which the operations are proposed to be directed are such as the avoidance of hail by the dissipation of thunder-clouds—this appeals particularly to the regions which surround the Alps. The production of rain in regions where rain is specially wanted for the maintenance of crops is another object; and thirdly, the dissipation of fog, and this last has now become transcendently important in flying.

The methods proposed are either mechanical or electrical.

Control of hail-storms by noise

The mediaeval tradition that the noise of church-bells, and subsequently of fire-arms, had a potent influence upon thunder-storms has a remarkable history. It expresses itself periodically in the vine-growing districts of Europe. It was epidemic in a very severe form at the end of last century because somebody had devised a new gun or mortar; pointed upwards, the mortar discharged a vortex ring of smoke which could be seen to reach the clouds. So popular did it become that nothing would content the people concerned but the expenditure of large sums of public money on installing batteries of these mortars, and they increased in size until they attained a height of 40 ft., about as high as a London house. A conference was held to decide whether the beneficent influence of the cannonade had been demonstrated by experience, and a vote was taken from which it appeared that the voters arranged themselves in order of the latitude of their domicile. Southerners thought the influence proved, Northerners thought it disproved, and the intermediate people thought it doubtful. So at the urgent instance of the persons concerned the Italian Government undertook experiments for two years and entrusted them to the physicist and Senator, Blaserna. By the time one year's experiments were over the epidemic had passed away, and the greatest unwillingness was evinced towards further expenditure of public funds on the inquiry.

Control by electrical discharge

This was at once followed by proposals for setting up paragrèles in France in the form of tall structures carrying metallic points for the discharge of electricity to neutralise the electricity of the thunder-clouds. I am afraid the war drew a veil over the results. I have lost sight of the paragrèle industry. Its prospects were not really hopeful because there are so many arrangements which act automatically on similar lines, such as trees, kites and kite balloons. But the idea has appeared elsewhere to be applied for the purpose of producing rain.

Production of rain by gunfire

Another development, in a different direction, of the belief in the effect of noise is that great gunfiring, dynamite explosions or any powerful detonations produce rainfall. It draws its support largely from the fact that many battles have ended in, or been followed by downpours of rain. Historically battles are summer phenomena and doubtless many summer days of less momentous importance have closed with downpours of rain. So widespread is the general opinion that during the wet summer of 1910 the farmers of the South of England petitioned that the gun-practice of the Fleet in the Solent might be postponed until they had got in the crops, which were being ruined. I can give some good reasons for associating thunder-storms with Naval reviews in July; but I hesitate to adduce them in evidence because the storms extended from Portsmouth to London and seemed a "tall" order for saluting guns to discharge.

In countries where rain is scarce and sometimes deficient the pressure of stock-owners to have something done to convert clouds into rain may be very painful. I remember being visited by a lady who begged me to say that she might spend the few remaining pounds of a fortune made in sheep-farming and lost for want of rain, in buying rockets to send out to be fired at the recalcitrant clouds which threatened rain but failed to produce it. She had already consulted a pyrotechnist who confined himself to saying he had no actual experience of making rain, but that his rockets would certainly be as effective for the purpose as those of any other firm. A good deal of ink and paper have been expended over arguing about the effects of gunfire and other explosions. It is difficult to argue about it effectively because there is no ground *a priori* for supposing that concussion or rockets would have any effect at all upon the condensation of vapour and clouds. And in any attempt to prove the influence, by rainfall which occurred subsequently to the explosions, we have no means of comparing actuality with what would have happened if the explosion had not occurred. So we have to rely on general reasoning. As to the rain that was associated with operations on the Western Front, the student of the weather maps of the time found the sequence of events always according to rule; he had no reason at all to suspect a local influence different from that of ordinary meteorological contingencies.

The effect of extensive gunfire may be regarded either as physical, arising from the detonation and thermal expansion, or chemical, due to the vast amount of material burned. The direct effect of the detonation is probably nothing at all; the thermal effect is insignificant compared with that of sunshine, and the chemical effect inconsiderable compared with the daily combustion of fuel in, say, the Manchester district.

Direct physical process

A certain Mr Cole and a Canadian airman are intending to protect regions of Canada from lack of rain by spraying liquid air from an aeroplane. That is quite a different story from the use of guns. They are certain to get the condensation corresponding with the reduction of temperature caused by the liquid air during its evaporation. It is only the experiment of the condensation of water on the outside of an ice-pitcher on a larger scale; but it seems unnecessary to carry the material to a great height and indeed there is a certain risk that the rain would evaporate before it reached the ground. A millimetre of rain means four tons to the acre, or 2500 tons to the square mile. To water a countryside would need a good deal of liquid air.

Control by throwing dust

Other suggestions for making rain are even less attractive. According to the *Daily Mail*, an attempt to produce rain by throwing dust from an aeroplane on to clouds 5000 ft. high in order to cause rain at Pretoria was unsuccessful, as well it might be, because the conditions in which a supply of dust might be effective in causing condensation, according to Aitken, is when clouds cannot form for want of nuclei. If the clouds are already there, the dust is certainly superfluous. An identical proposal was made to the Royal Meteorological Society some years before the war, when shovelling dust out of a balloon or aircraft over London was prescribed as a means of dissipating fog.

Production of rain by electrical discharge

The effect which has been alleged as following gunfire has also been claimed for electrical operations. An electrical installation in Australia for discharging electricity from kites was said to have produced enough rain to fill a large tank in a region that suffered from lack of rain; the ordinary meteorological observations of the time showed that not only the particular locality was affected but the whole country for hundreds of miles round was uniformly fortunate. And it appears to be a question of psychology whether you regard the general weather conditions of a continent as being affected by the local installation, or independent of it.

So we approach the practical question of the control of our weather with the consciousness that not even experience is allowed to be conclusive by both sides, and the views arrived at express psychology rather than pure science.

Scale of operations in the open air

For one thing, these questions are questions of scale. We can do anything with a quantity of air in a small enclosure in a laboratory. We can certainly, by artificial means, make cloud or rain in the enclosure, and disperse it or evaporate it at will after it has been formed. We could easily find out whether the detonation of a pistol or a small charge of dynamite at a suitable distance would produce any effect upon an artificial cloud, though I have never heard of the experiment being tried. The important question is whether we can extend such operations from the laboratory to the open air.

We are here up against the important consideration that a cube of air, 10 metres (about 33 ft.) each way, weighs more than a ton. If it is foggy it may contain 5 kilogrammes of water drops, and a millimetre of rain over the same area (about 120 square yards) weighs 100 kilogrammes. The amount of heat released by the condensation of a kilogramme of water is about 600 kilogramme-centigrade units, which is equivalent to $2 \cdot 5 \times 10^{13}$ ergs, or, approximately, 1 horsepower-hour ($2 \cdot 7 \times 10^{13}$ ergs). Hence evaporating fog in a 10-metre cube of air is equivalent to 5 horsepower-hours, and a millimetre of rain—over the 10-metre square—represents energy to the extent of 100 horsepower-hours; over a square kilometre, a million horsepower-hours. Amounts of energy in these proportions have to be disposed of, or developed respectively, when the corresponding condensation is caused or reversed. With increase of scale, the amounts of energy involved soon pass beyond the limits of human control.

In order to give you a definite idea of the kind of effect which great dimensions have upon the prospects of human control, let me adduce a simple case in which dimensions are easily realised. A suggestion was made some years ago to protect the steamer routes of the Atlantic by diverting the Labrador Current. The project involved the building of a jetty 200 miles long from Newfoundland. There is nothing which can be called impossible about building a jetty a mile long, and 200 miles is only two hundred times as long. It can only be a question of money, material and perseverance. But for practical purposes, impossibility is reached when the money and material required exceed the limit of what is available, and it is from that point of view that all proposals for the human control of weather have to be viewed.

What the Atlantic Ocean would have to say to a two-hundred-mile jetty when it was built or while it was building is another matter. It might distribute the material in a manner which differed from the specification. While, therefore, one cannot say that such an enterprise is impossible, it is not attractive in these hard times.

I have another proposal of a different character: this time to arrest and prevent the development of fog at sea by pouring oil on to the water and so stop evaporation in the environment of the ship. In this case, it is not merely the scale; the basic theory is probably at fault. The water of an Atlantic fog does not, as a rule, come from the surface on which the fog lies, but from far to the south. It is the cold surface which causes the fog; the temperature of the surface is below the dew-point of the air above it, and dew would therefore be formed on the oil. Even if the theory were correct and we obtained a patch of oil, a clear space, and a ship, we should still have to consider what would be their relative positions at the end of an hour or twelve hours, in view of their relative drifts. An identical method was suggested some years ago for application to the river Rhone, at its junction with the Saône, where warm and cold water join. No news has arrived as to the success of the proposal.

Clearance of fog from aerodromes

With these examples, let us turn to the modern problem of clearing fog from aerodromes. It would be a work of the most obvious utility, and even urgency, if it could be accomplished. You will permit me to pass over with mere mention a proposal which was within a little of being adopted some twelve years ago by the London County Council, to use for the purpose of dispersing the fog of London the mortars which were originally designed to convert the destructive hail-storms of Italy into beneficent rain. Instead of getting £5000 wherewith to try the experiment, the promoters only got the promise of the use of sites for placing their batteries, and the trial was never made. I cannot avoid the conclusion that this suggestion arose not from any scientific observation of the effect of detonations or smoke rings upon fog, but from the fact that the mortar was designed as an engine for the human control of weather; the control required in London was for fogs and not thunder-storms, and anyway London could afford to pay. But for saying so I got severely criticised at the time in one newspaper.

The effect of wind

It would appear from experience that the easiest way of disposing of the comparatively calm fogs of an aerodrome is to get up a slight wind and blow them away. Captain Carpenter, in his report on London fogs in 1902, found that valley fogs could not survive a record of wind at Kew beyond 13 m.p.h.

(Factor 3); the same rule would not work for hill-tops. Such a wind corresponds with a difference of pressure of a millibar in 75 nautical miles. A bank of air three metres high along one side of a quarter-mile aerodrome would be sufficient, and it seems rather absurd to call the maintenance of such a bank impossible. But it is so.

Dissipation of fog by heat

A more reasonable suggestion was made to me some months ago in a letter from a flying officer. He had noticed that the players in a football match which he was watching kept themselves clearly visible, while the rest of the ground was befogged up to a thickness of about 50 feet. He supposed that the air was dead calm, and spaces might therefore be permanently cleared by local heating. It is, however, an essential peculiarity of fog that the air in which it floats is never really still; it always has a slow drift, as anyone can see who watches a fog from the inside. In fact, if there were no drift there would be no fog problem: the drops would sink to the ground. Gravity would do the work of removing them in the simplest possible manner. It is only the eddy-motion accompanying the drift that keeps the drops persistently in the air by preventing them settling. Taking the drift at two miles an hour, I made a rough computation of the coal required to clear an aerodrome 400 yards wide. It worked out at about 12 tons an hour for coal consumption for a 50 ft. fog, and ran up to 400 or 500 tons an hour, as an outside figure, to meet ordinary contingencies, using electrical distribution. Again, it is simply a question of magnitude. I have myself no practical conception of the amount of combustion which is implied by 12 tons or 100 tons an hour. My sheet-anchor about coal is that a fire in my college room used about 2 cwt. in a week of about 100 hours, or about one-thousandth of a ton per hour. So 12 tons per hour is the equivalent of 12,000 college rooms; shall we say, five times the consumption of the University and Colleges of Cambridge? Such an amount of combustion is hardly to be called impossible, but no other adjective is so nearly an expression of the facts.

My feeling about attempting such experiments is perhaps best described by saying that the problem is about the same as trying to raise by a few degrees the temperature of the top 2 ins. of the Thames between the Lots Road Power Station and Battersea Bridge when the tide has just begun to ebb. I would not like to say it is impossible with unlimited funds and coal. I do not know how much coal they burn in an hour at Lots Road, but if the plan is to be tried, it had better be on a small brook first.

Mechanical drainage of air

If we approach the same problem by mechanical means and endeavour to drive away the foggy air of an aerodrome by propellers capable of giving a speed of 100 kilometres per hour to the propelled stream, we find ourselves in the same difficulty. We arrive at figures for which "impossible" is only too strong a word if you disregard all questions of cost and effort.

ELECTRICAL PROSPECTS   225

The unexplored electrical force

But in these days we not only have the advantage of dynamical contrivances like the internal combustion engine, which packs so much available energy into so small a compass, but we are entering upon a new kingdom of electrical action which is as yet very imperfectly explored so far as the atmosphere is concerned. I have certainly heard both Sir Oliver Lodge and the Master of Trinity claim that it is possible to affect the weather by electrical operations; and one of them, I forget which, has made a certain amount of play with the disputes which will arise between neighbours in the endeavour to obtain control of the machine in their own interest. But I do not know what their practical proposals are for the direct control of the weather, so we have to go back to facts.

Once more we know that on the small scale of an enclosure within a laboratory a brush discharge of electricity will clear away dust, smoke and cloud like a magic. We know that the process has been extended to the larger scale of furnace flues. There is already a company incorporated to construct suitable apparatus for clearing such flues, and the only question is whether the same process would be operative in the free air to a sufficient extent to clear an aerodrome of fog. We have, to guide us, only the records of the experience of Sir Oliver Lodge, who erected on the roof of his laboratory in Liverpool a discharging conductor for the purpose of clearing Liverpool from fog. It was an object of curious interest to all passers-by for many years. Rumour has it that it was brought into operation on one occasion, and on that occasion the space round the laboratory became clear of fog. At the same time, or nearly so, the fog cleared away from the whole of Liverpool, and, as far as I know, the experiment has never been repeated. You must form your own opinion whether it was the weather or the electricity that cleared the neighbourhood of the laboratory.

Until further experiments throw new light on the subject, I think the betting is on the weather, because the operation of clearing away dust by electrical action seems to be dependent upon the brush discharge and not on the steady current of electricity carried by ions in the atmosphere. A brush discharge comes pretty near to sparking, and to make an electrical installation that is within range of sparking across an aerodrome is, if not impossible, at least a very serious matter from many points of view. We must remember that already the drops of water in the air are subject to the separating force represented by the differential effect of gravity upon the drops and the air which carries them, and therefore the force which is necessary to drive drops through the air to electrodes on either side must be large compared with the force of gravity upon the drops; that force which is already operative all the time produces no apparent effective result in consequence of the counteracting effect of eddy-motion. At the moment, though not hopeful, I keep an open mind upon the possibility of clearing a space on an aerodrome by electric discharge. If you ask me for an opinion, I shall ask to put a question to you in return arising out of a suggestion that came before me during the war as a means of producing disagreeable weather for our enemies. It was to create a tornado by firing shells vertically upward in such rapid succession that they would produce a vortex with a vertical axis. Now I am particularly interested just now in the formation of tornadoes, so I shall leave this question with you: How many guns would you have to fire, and how frequently, in order to produce a tornado in the way suggested?

It is, I believe, so far analogous to the electrical question in that, apart from the ultimate destinations of the shells, it depends upon scale. The most telling example of malevolence of the weather towards the Allied Forces that I can recall in the course of the war is the development of a rainy cyclonic depression over the Western Front and southern part of the North Sea during the end of July and the beginning of August, 1917. It began to form on July 28, and reached its climax on August 3, when a well-marked depression, 11 millibars deep, was exhibited on the map, extending over a nearly circular area 1400 kilometres in diameter, and had filled up on August 6. It apparently originated and filled up again in the locality. I reckon that the creation of the depression, which was a very small affair, and on the map looked like gerry-mandering, is equivalent to the removal from within the cylinder of 1400 kilo-metres diameter of seventy thousand million tons of air. It took six days to accomplish this deportation, and three days to fill the space up again. If the enemy accomplished this feat by artificial means, they must have used some other process than firing shells vertically upwards: the question gives me the same sort of tired feeling as the 200-mile jetty, with some other sensations added.

The most direct means of accomplishing such a deportation of air would be by an underground channel to carry the air from the central region to beyond the boundary of the depression. Let us suppose a channel, twelve feet in diameter, leading from Ostend to Berlin and operated there by a 16 ft. propeller giving a full bore stream of 100 kilometres an hour (friction being neglected). The deportation would go on at the rate of 1200 tons per hour, or 28,800 tons per day. Working without intermission, it would take 7000 years for the propeller to complete the deportation; and as it had to be done in six days, 400,000 such channels would have to be operated concurrently to get the work done in time. If the Germans had started drawing air out when they began to speak German at the time of the tower of Babel they would just have got the work completed in time for August 3, 1917.

The expenditure of energy

What it comes to, then, is that all the suggestions for the human control of weather oppress one, not always by mistaken conception of physical processes, but by the "scale effect." Within our knowledge we are lords of every single specimen of the atmosphere which we can bottle up and imprison in our laboratories, our furnace flues, or our greenhouses; but in the open air the ordinary inexorable laws which control the behaviour of the atmosphere, when we are awake and when we are asleep, have such enormous quantities of energy in the form of warmth and water-vapour in reserve that our own little reserves are not equal to making any serious impression on the course of nature.

Yet the course of the weather may be affected by what may be regarded as violent artificial means, such as the explosions of a great volcano. In a recent work by Professor W. J. Humphreys the suggestion has been put forward that cold summers and even glacial periods have been caused in that way, and I see a prize is offered for an essay on the connection between vulcanism and storms, among other things.

So perhaps we might give a new turn to our thoughts by exploring how far our reserves of available energy compare with the destruction of Pompeii, the disappearance of the island of Krakatoa, of the eruptions of Mont Pelée and la Souffrière. In any case, it is the law of conservation of energy which we have to bear in mind, and it is the vastness of the volume and mass of the air affected which has hitherto offered insuperable obstacles to the application of known physical processes for the control of weather. Any new physical process, to be successful, will have to arrange for a great economy in the energy required or give us access to supplies of energy which are not now available.

APPENDIX

ACCELERATION OF THE AIR BY FALLING RAIN
(See pages 81 and 90.)

NOTE BY DR S. FUJIWHARA, OCTOBER, 1921

Notation:

m, mass of raindrops in 1 m³ of air.

M, mass of 1 m³ of air.

g, acceleration of gravity.

a,   ,,   ,,   raindrops.

A,   ,,   ,,   air.

v, velocity of falling raindrops.

V, induced velocity of the air.

Assumptions (first approximation):

1. Raindrops start from rest (this is not important, the drops can start from any velocity without changing the result).

   Air mass M is constant.

   Rainfall is continuous and the height great enough for the drops to reach a steady velocity.

2. The frictional force F between the air and raindrops is proportional to the difference of velocities, i.e. $F = k\,(v - V)$.

   There is a limit $(v - V)$ at which the frictional force balances the weight of the drops (assumed of uniform size), then $k\,(v - V)_0 = mg$.

   The acceleration of the raindrops is zero when $(v - V) = (v - V)_0$ and is g when $(v - V) = 0$.

We shall therefore assume:

$$\frac{dv}{dt} = \frac{(v - V)_0 - (v - V)}{(v - V)_0}\, g,$$

$$M\frac{dV}{dt} = k\,(v - V).$$

At the limiting velocity $\dfrac{dV}{dt}$ will be $\dfrac{m}{M}g$, hence $k = m\,\dfrac{g}{(v - V)_0}$,

$$\frac{dV}{dt} = \frac{m}{M}\frac{g}{(v - V)_0}\,(v - V) \quad\dotfill\text{(A)},$$

$$\frac{dv}{dt} = \frac{(v - V)_0 - (v - V)}{(v - V)_0}\, g \quad\dotfill\text{(B)}.$$

The total force on raindrops and air-mass is constant, hence:

$$M\frac{dV}{dt} + m\frac{dv}{dt} = mg \quad \text{......................(C)};$$

from this also it follows that $k = m\dfrac{g}{(v - V)_0}$.

Subtracting (A) from (B)

$$\frac{d(v - V)}{dt} = g - \frac{v - V}{(v - V)_0}g - \frac{m}{M}\frac{g}{(v - V)_0}(v - V)$$

$$= g - \frac{g}{(v - V)_0}\left(1 + \frac{m}{M}\right)(v - V)$$

$$= g - g\frac{m + M}{M}\frac{v - V}{(v - V)_0},$$

$$d(v - V) + g\left(\frac{m + M}{M}\right)\frac{v - V}{(v - V)_0}\,dt = g\,dt.$$

Multiplying by $e^{\frac{g}{(v - V)_0}\frac{m + M}{M}t}$:

$$e^{\frac{g}{(v-V)_0}\frac{m + M}{M}t}\left[d(v - V) + \frac{v - V}{(v - V)_0}\left(\frac{m + M}{M}\right)g\,dt\right] = ge^{\frac{g}{(v-V)_0}\frac{m + M}{M}t}\,dt.$$

Integrating we get:

$$v - V = e^{-\frac{g}{(v-V)_0}\frac{m+M}{M}t}\left(\int ge^{\frac{g}{(v-V)_0}\frac{m+M}{M}t}\,dt + C\right)$$

$$= Ce^{-\frac{g}{(v-V)_0}\frac{m+M}{M}t} + \frac{M}{m + M}(v - V)_0.$$

When $t = 0$, $v = V = 0$.

$$\therefore C = -(v - V)_0\frac{M}{m + M}.$$

Hence: $$\frac{v - V}{(v - V)_0} = \frac{M}{m + M}\left(1 - e^{-\frac{m + M}{M}\frac{g}{(v - V)_0}t}\right).$$

From equation (A)

$$\frac{dV}{dt} = \frac{mg}{M + m}\left(1 - e^{-\frac{m + M}{M}\frac{g}{(v - V)_0}t}\right),$$

$$V = \frac{mg}{M + m}\left(t + \frac{M}{m + M}\frac{(v - V)_0}{g}e^{-\frac{m + M}{M}\frac{g}{(v - V)_0}t}\right) + C'.$$

When $t = 0$, $V = 0$, hence $C' = -\dfrac{mM}{(m + M)^2}(v - V)_0$.

Whence:

$$V = \frac{mg}{M + m}t - \frac{mM}{(m + M)^2}(v - V)_0\left(1 - e^{-\frac{m + M}{M}\frac{g}{(v - V)_0}t}\right),$$

and $$v = \frac{mg}{M + m}t + \frac{M^2}{(m + M)^2}(v - V)_0\left(1 - e^{-\frac{m + M}{M}\frac{g}{(v - V)_0}t}\right).$$

Actual Calculations:

Assume
$$(v - V)_0 = 8 \text{ m/s},$$
$$m = 1 \cdot 25 \text{ g/m}^3,$$
$$M = 1250 \text{ g/m}^3.$$
$$V = 0 \cdot 98t - \tfrac{800}{1000} (1 - e^{-1 \cdot 225t}).$$
$$v = 0 \cdot 98t + 800 (1 - e^{-1 \cdot 225t}).$$

Height of cloud (metres)...	1000	2000	3000	4000
Time required (secs.) ...	120	240	375	500
v at the end of time t (m/s)	9·17	10·35	11·7	12·9
V „ „ „	1·2	2·3	3·7	4·9

*   *   *

Taking the fall from 2000 metres $V = 2 \cdot 3$ m/s and supposing the air brought down is dispersed over a circle radius r and thickness h, the radius of the shower being also r, and $V'/2$ the mean velocity of outflow,
$$\pi r^2 V = 2\pi r h V'/2,$$
$$rV = hV', \text{ where } V' \text{ is the horizontal surface velocity.}$$
If $r = 500$ m. and $h = 50$ m.
$$V' = 23 \text{ m/s}.$$

The estimate of rain, viz. $1 \cdot 25$ g/m³ falling at the maximum relative rate $v - V = 8$ m/s would give drops falling at $10 \cdot 35$ m/s and therefore
$$10 \cdot 35 \times 1 \cdot 25 \text{ g/m}^2.$$
$$12 \cdot 93 \text{ g/m}^2 \text{ per sec.} = \frac{12 \cdot 93 \times 10}{10000} \text{ mm. depth of rain per sec.}$$
$$= \frac{6 \times 1293}{10000} \text{ mm/min.}$$

This corresponds with a fall of $46 \cdot 5$ mm. of rain per hour.

*   *   *

Note, June, 1922: Professor W. J. Humphreys has recently pointed out that the condensation of water-vapour into rain in a vertical column can make no difference to the air-pressure at the bottom because the mass remains always the same. It is, however, not likely that the reasoning would apply to the free atmosphere because the limitation to a vertical column of unchanging material is not satisfied. Wherever the air may be when condensation is in progress, the position in vertical column over the rain which it has produced is the most unlikely of all the possible hypothetical positions.

N. S.

INDEX

www.ingramcontent.com/pod-product-compliance
Ingram Content Group UK Ltd.
Pitfield, Milton Keynes, MK11 3LW, UK
UKHW050116180125

453697UK00014B/447